ELIMINATING
HEALTH DISPARITIES
MEASUREMENT AND DATA NEEDS

Panel on DHHS Collection of Race and Ethnicity Data
Michele Ver Ploeg and Edward Perrin, Editors

Committee on National Statistics
Division of Behavioral and Social Sciences and Education

NATIONAL RESEARCH COUNCIL
OF THE NATIONAL ACADEMIES

THE NATIONAL ACADEMIES PRESS
Washington, D.C.
www.nap.edu

NATIONAL ACADEMIES PRESS 500 Fifth Street, NW Washington, DC 20001

NOTICE: The project that is the subject of this report was approved by the Governing Board of the National Research Council, whose members are drawn from the councils of the National Academy of Sciences, the National Academy of Engineering, and the Institute of Medicine. The members of the committee responsible for the report were chosen for their special competences and with regard for appropriate balance.

This study was supported by Contract/Grant No. HHS-100-01-0022 between the National Academy of Sciences and the U.S. Department of Health and Human Services. Any opinions, findings, conclusions, or recommendations expressed in this publication are those of the author(s) and do not necessarily reflect the views of the organizations or agencies that provided support for the project.

Library of Congress Cataloging-in-Publication Data

Eliminating health disparities : measurement and data needs / Panel on DHHS Collection of Race and Ethnicity Data ; Michele Ver Ploeg and Edward Perrin, editors.
p. ; cm.
Includes bibliographical references.
ISBN 0-309-09231-0 (pbk.)
1. Social medicine—United States—Methodology. 2. Health status indicators—United States—Measurement. 3. Social indicators—United States—Measurement. 4. Economic indicators—United States—Measurement. 5. Medical care—United States—Evaluation—Statistical methods. 6. Health services accessibility—United States—Evaluation—Statistical methods. 7. Minorities—Health and hygiene—United States—Statistical methods. 8. Ethnic groups—Health and hygiene—United States—Statistical methods.
  [DNLM: 1. Data Collection—United States. 2. Ethnic Groups—United States. 3. Evaluation Studies—United States. 4. Health Status Indicators—United States. 5. Socioeconomic Factors—United States. 6. Treatment Outcome—United States. WA 300 E424 2004] I. Ver Ploeg, Michele. II. Perrin, Edward. III. National Research Council (U.S.). Panel on DHHS Collection of Race and Ethnicity Data.
  RA418.3.U6E45 2004
  362.1′0973—dc22

                              2004012734

Additional copies of this report are available from National Academies Press, 500 Fifth Street, NW, Lockbox 285, Washington, DC 20055; (800) 624-6242 or (202) 334-3313 (in the Washington metropolitan area); Internet, http://www.nap.edu

Suggested citation: National Research Council. (2004). *Eliminating Health Disparities: Measurement and Data Needs.* Panel on DHHS Collection of Race and Ethnicity Data, Michele Ver Ploeg and Edward Perrin, Editors. Committee on National Statistics, Division of Behavioral and Social Sciences and Education. Washington, DC: The National Academies Press.

THE NATIONAL ACADEMIES
Advisers to the Nation on Science, Engineering, and Medicine

The **National Academy of Sciences** is a private, nonprofit, self-perpetuating society of distinguished scholars engaged in scientific and engineering research, dedicated to the furtherance of science and technology and to their use for the general welfare. Upon the authority of the charter granted to it by the Congress in 1863, the Academy has a mandate that requires it to advise the federal government on scientific and technical matters. Dr. Bruce M. Alberts is president of the National Academy of Sciences.

The **National Academy of Engineering** was established in 1964, under the charter of the National Academy of Sciences, as a parallel organization of outstanding engineers. It is autonomous in its administration and in the selection of its members, sharing with the National Academy of Sciences the responsibility for advising the federal government. The National Academy of Engineering also sponsors engineering programs aimed at meeting national needs, encourages education and research, and recognizes the superior achievements of engineers. Dr. Wm. A. Wulf is president of the National Academy of Engineering.

The **Institute of Medicine** was established in 1970 by the National Academy of Sciences to secure the services of eminent members of appropriate professions in the examination of policy matters pertaining to the health of the public. The Institute acts under the responsibility given to the National Academy of Sciences by its congressional charter to be an adviser to the federal government and, upon its own initiative, to identify issues of medical care, research, and education. Dr. Harvey V. Fineberg is president of the Institute of Medicine.

The **National Research Council** was organized by the National Academy of Sciences in 1916 to associate the broad community of science and technology with the Academy's purposes of furthering knowledge and advising the federal government. Functioning in accordance with general policies determined by the Academy, the Council has become the principal operating agency of both the National Academy of Sciences and the National Academy of Engineering in providing services to the government, the public, and the scientific and engineering communities. The Council is administered jointly by both Academies and the Institute of Medicine. Dr. Bruce M. Alberts and Dr. Wm. A. Wulf are chair and vice chair, respectively, of the National Research Council.

www.national-academies.org

Acknowledgments

I would like to thank, on behalf of the Panel on DHHS Collection of Race and Ethnicity Data, all of the individuals involved in the production of this report. I first would like to thank our sponsors within DHHS, the Office of the Assistant Secretary for Planning and Evaluation, the Agency for Healthcare Research and Quality, the Centers for Disease Control and Prevention, the Health Resources and Services Administration, the HIV/AIDS Contigency Fund, the National Institutes of Health, the Office for Civil Rights, and the Office of Minority Health. We would like to especially thank James Scanlon and Dale Hitchcock of the Office of the Assistant Secretary for Planning and Evaluation (ASPE) for their continual assistance as liaisons for the panel to DHHS. We would also like to thank the staff of the Centers for Medicaid and Medicare Service and of the Social Security Administration who met with National Research Council's Committee on National Statistics (CNSTAT) staff regarding racial and ethnic data collection for these agencies' programs.

Many individuals gave presentations to the panel on various data related topics and should be thanked for taking time out of their busy schedules. Vickie Mays of the University of California Los Angeles briefed the panel on the data activities of the National Committee on Vital and Health Statistics, Subcommittee on Special Populations. William Braithwaite, of PriceWaterhouseCoopers, provided information on the data collection requirements of the Health Insurance Portability and Accountability Act. The panel is also indebted to those individuals who wrote background papers for this report and for the panel's Workshop on Improving Racial and

Ethnic Data in Health. Gratefully we thank Patricia O'Campo, Jessica Burke, Allen Fremont, Nicole Lurie, Jeffrey Geppert, Sara Singer, Jay Buechner, Lorin Ranbom, Walter Suarez, Wu Xu, David Nerenz, Connie Currier, and Carmella Bocchino. The papers written by these individuals appear in the Appendixes to this report.

The panel is grateful for the excellent work of the staff of CNSTAT and the National Research Council for developing and organizing the workshop and for writing this report. Michele Ver Ploeg, study director for the panel and coeditor of the report, should be thanked for steering the panel throughout its activities. She drafted major sections of the report and guided the report through the review process. The panel would like thank to Jamie Casey, research assistant, for providing excellent service to the panel—diligently assembling background materials and tables and figures for the report. The panel is also thankful for the efforts of Tanya Lee, project assistant, in handling all administrative matters regarding the panel and in the production of this report. Earl Pollack, program officer, read through early drafts and made valuable contributions to the editing and polishing of the report. The panel was also ably aided by the services of two consultants. Mary Grace Kovar met with several Department of Health and Human Services staff members regarding the details of their data systems, and Daniel Melnick drafted the summary of the Workshop on Improving Racial and Ethnic Data in Health (released in 2003) and commented on early drafts of the panel's final report. Kirsten Sampson Snyder and Chris McShane of the reports office of the Division of Behavioral and Social Science Education, and Cameron Fletcher, are thanked for guiding the workshop report through the review process and for professional editing of the report.

As chair of this panel, I thank my fellow panel members for giving their time and expertise so generously toward the completion of this report. Each of their contributions to the discussions in the development and to the drafting of this report is greatly appreciated.

This report has been reviewed in draft form by individuals chosen for their diverse perspectives and technical expertise, in accordance with procedures approved by the Report Review Committee of the National Research Council. The purpose of this independent review is to provide candid and critical comments that will assist the institution in making its published report as sound as possible and to ensure that the report meets institutional standards for objectivity, evidence, and responsiveness to the study charge. The review comments and draft manuscript remain confidential to protect the integrity of the deliberative process. We thank the following individuals for their review of this report: Olivia Carter-Pokras, Department of Epidemiology and Preventive Medicine, University of Maryland School of Medicine; Sheldon Greenfield, Department of Medicine, University of California, Irvine; Judith R. Lave, Department of Health Policy and Management, Graduate School of Public Health, University of Pittsburgh; Vickie M. Mays, Department of Clinical Psychology, University of California, Los Angeles; Dorothy P. Rice, Department of Social and Behavioral Sciences Emeritus, School of Nursing, University of California, San Francisco; Robert Santos, Executive Office, NuStats, Austin, TX; Kenneth E. Thorpe, Department of Health Policy and Management, Rollins School of Public Health, Emory University; and Clyde Tucker, Office of Survey Methods Research, Bureau of Labor Standards, Washington, DC.

Although the reviewers listed above have provided many constructive comments and suggestions, they were not asked to endorse the conclusions or recommendations nor did they see the final draft of the report before its release. The review of this report was overseen by Richard A. Kulka, Social and Statistical Sciences, Research Triangle Institute. Appointed by the National Research Council, he was responsible for making certain that an independent examination of this report was carried out in accordance with institutional procedures and that all review comments were carefully considered. Responsibility for the final content of this report rests entirely with the authoring committee and the institution.

Edward Perrin, *Chair*
Panel on DHHS Collection of Race and
Ethnicity Data

Contents

APPENDIXES

ELIMINATING
HEALTH DISPARITIES
Measurement and Data Needs

Executive Summary

Disparities in health and health care across racial, ethnic, and socioeconomic backgrounds in the United States are well documented. The reasons for these disparities are, however, not well understood. Considerable interest in better understanding the causes of these differences has called attention to the availability and quality of individual-level data on race, ethnicity, socioeconomic position (SEP) and acculturation and language (e.g., language use, place of birth, generation status) of individuals. These data are critical to documenting the nature of disparities in health care and to developing strategies to eliminate disparities.

Data currently available on race, ethnicity, SEP, and acculturation and language use are severely limited. While national-level surveys sponsored or conducted by the federal government collect rich information on individuals, their health, and their use of health care, sample sizes often limit their usefulness to only broad racial and ethnic groups (e.g., blacks, whites, and Hispanics) and are typically too small for analyses within racial and ethnic groups (e.g., within the Hispanic ethnic category—Puerto Ricans, Cubans, Mexicans, and other Hispanic groups) or for smaller, but still broad, racial and ethnic groups (e.g., American Indians and Alaska Natives). Data from Medicare claims and enrollment files have been widely used for analysis of racial and ethnic disparities. However, racial and ethnic data in these files are of limited accuracy, completeness, and detail. State-based data, such as vital records, administrative data from Medicaid and the State Children's Health Insurance Program, and data from registry systems, are potentially valuable sources of data for analyzing disparities in health and health care.

However, these data sources do not collect data on race and ethnicity in standardized ways, and they contain little information on other relevant patient characteristics. Finally, although much information on health and health care comes from private data systems maintained by health insurance plans, hospitals, and medical groups, data on race and ethnicity usually are not collected in these record systems. When the information is available, it is often unstandardized and contains little information on patients' socioeconomic characteristics or acculturation and language use. The lack of standardized and complete data challenges the establishment of reliable baseline and trend analyses of health, health care access, cost, and quality by patient characteristics.

Concerns about the adequacy of the current infrastructure to provide the necessary data to understand and eliminate racial and ethnic disparities prompted Congress to direct the Department of Health and Human Services (DHHS) to request that the National Academies conduct a comprehensive study of DHHS data collection systems (P.L. 106-525, 2000). In response to this request, the DHHS Office of the Assistant Secretary for Planning and Evaluation (ASPE), on behalf of a number of agencies within DHHS, asked the Committee on National Statistics (CNSTAT) of the National Academies to convene a panel of experts to review DHHS data systems.[1] ASPE and CNSTAT developed the charge for the study based on this legislation and on the department's own needs for review of its data systems, giving the panel the flexibility to review related data needs as they arose.

The panel was charged to review data collection or reporting systems required under the department's programs or activities relating to the collection of data on race, ethnicity, and socioeconomic position. The charge included examining data collection systems in other federal agencies with which the department interacts to collect relevant data on race and ethnicity (such as that of the Social Security Administration [SSA]), as well as systems of the private health care sector. The panel was asked: (1) to identify the data needed to support efforts to evaluate the effects of socioeconomic position, race, and ethnicity on access to health care and on disparities in health as well as to enforce existing protections for equal access to heath care; (2) to assess the effectiveness of the data systems and collection practices of DHHS and of selected systems and practices of other federal, state, and local agencies and the private sector in collecting and analyzing such

---

[1]Other DHHS agency sponsors include the Agency for Healthcare Research and Quality (AHRQ), Centers for Disease Control and Prevention (CDC), Health Resources and Services Administration (HRSA), HIV/AIDS Contingency Fund, National Institutes of Health (NIH), Office for Civil Rights (OCR), and Office of Minority Health (OMH).

data; and (3) to identify critical gaps in the data and suggest ways in which they could be filled.

We note some specific distinctions the panel made in interpreting its charge. First, the panel reviewed a very broad set of data collection systems both within and outside DHHS. These systems include health surveys, administrative records, and records from private data systems. The research purposes and uses of these data collection systems are quite varied—some are used to understand broad determinants of health (e.g., the effect of income on mortality) while others are used to understand very specific outcomes of health care treatment (e.g., the effects of ethnicity and race on medical outcomes of patients with hypertension or diabetes). The panel focused only on the collection of data on race, ethnicity, and socioeconomic position (as the originating legislation called for), and added to that the collection of data on acculturation and language use because the panel believed these to be important correlates to understanding racial, ethnic, and socioeconomic aspects of health and health care. In making recommendations, the panel did not consider specific assessments of the cost of improved data collection but did broadly consider the costs of data collection among different types of data collection systems.

THE IMPORTANCE OF DATA ON RACE, ETHNICITY, SOCIOECONOMIC POSITION, AND ACCULTURATION AND LANGUAGE USE

High-quality data on race and ethnicity are necessary to identify and eliminate disparities in health and health care. Socioeconomic position (SEP)—income, wealth, and education—is important as both a mediator of racial and ethnic disparities and a further source of disparities. Low SEP, for example, is associated with limited access to the health care system, inadequate health information, and poor health practices. Acculturation (and its proxy measures language, place of birth, years in the United States, or generational status) is also related to health status; mismatches between the language spoken by health care providers and by patients can be a limiting factor in health care interactions and health information exchange. The panel therefore concluded that:

CONCLUSION 3-1: Measures of race and ethnicity should be obtained in all health and health care data systems.

CONCLUSION 3-2: Measures of socioeconomic position should, where feasible, be obtained along with data on race and ethnicity.

CONCLUSION 3-3: Measures of acculturation and proxies such as language use, place of birth, and generation and time in the United States should, where feasible, be obtained.

To monitor trends in disparities, to understand how disparities arise, and ultimately to design interventions to eliminate and reduce them, information about individuals is used to make general statistical inferences about populations. Such statistical uses are distinct from other uses of the data that require information about a specific individual. For example, income data on individuals applying for Medicaid are collected to assess eligibility for the program. In this example, data on individuals are collected to take action regarding a specific individual. In contrast, data on individuals used for statistical purposes are collected to make inferences at an aggregate level.

Many of the data used to understand health disparities are not collected specifically for these statistical purposes, but rather are used to administer services and programs. Their use for statistical purposes is secondary. The panel, in this report, will make recommendations that encourage the collection of additional items of race and ethnicity, SEP, and language and acculturation where possible so that statistical inferences about disparities can be made. But it does so with the recognition that these data need to be useful to the federal, state, and private institutions and systems for which they are collected.

CONCLUSION 3-4: Health and health care data collection systems should return useful information to the institutions and local and state government units that provide the data.

Data linkages, or bringing together variables from two or more data sets, can facilitate new analyses (for policymaking, quality improvement, and research) without the expense and time needed for additional data collection. While there are tremendous opportunities for new analyses with linked data, barriers to data linkage—confidentiality concerns and negotiating linkages across different agencies each with their own protection rules, for example—are substantial. However, methods such as masking and deidentification can be used to guard against harmful uses of linked data and to protect confidentiality. Linking across data sets has great potential payoff in terms of increased content coverage over a single source of data.

CONCLUSION 3-5: Linkages of data should be used whenever possible, with due regard to proper use and the protection of confidentiality in order to make the best use of existing data without the burden of new data collection.

DHHS DATA COLLECTION SYSTEMS

In its evaluation of gaps in the department's data collection systems, the panel reviewed the 1999 DHHS report *Improving the Collection and Use of Racial and Ethnic Data in Health and Human Services*, a comprehensive study of the federal issues related to racial and ethnic data collection. The report calls for DHHS to develop an implementation plan that would prioritize the report's recommendations, include a detailed plan of action, establish a responsible office(s) to carry out the plan, and assess costs for implementation. Thus far, such a plan has not been produced. The panel urges DHHS to develop such a plan, begin to implement the data improvement recommendations, and establish a responsible body for coordinating implementation across the various department agencies and ensuring that they follow through with recommendations.

RECOMMENDATION 4-1: DHHS should begin immediately to implement the recommendations contained in the 1999 report entitled *Improving the Collection and Use of Racial and Ethnic Data in Health and Human Services.*

There are many important recommendations in the 1999 DHHS report. However, the panel emphasizes a few that it sees as priorities for the department.

National household surveys are not large enough to support analysis of health outcomes for many racial and ethnic subgroups. In addition, the costs of obtaining extensive health data that are collected in surveys like the National Health Interview Survey (NHIS) or the National Health and Nutrition Examination Survey (NHANES) for small or geographically concentrated racial and ethnic groups make it impossible to collect such data on a regular basis for every group that may be of interest. However, periodic targeted studies on specific groups in specific areas could be conducted and could provide essential data on the health of these groups. The panel therefore recommends that DHHS develop a schedule for special targeted surveys of population subgroups, covering a 10- to 20-year period and each year identifying the group to be targeted. This would be a feasible way of collecting meaningful data on racial and ethnic subgroups over time.

RECOMMENDATION 4-2: DHHS should conduct the necessary methodological research and develop and implement a long-range plan for the national surveys to periodically conduct targeted surveys of racial and ethnic subgroups.

Beyond sample size, there may be other statistical issues to address when surveying certain racial and ethnic groups such as recent immigrants and farm workers (Kalsbeek, 2003). The rarity of these groups, their geographic dispersion, and in some cases their mobility often make it ineffi-cient to sample them using standard household sampling methods. Further-more, survey questions might be understood differently by different groups. These factors can distort measures of disparities. For these reasons, special methods are needed to measure disparities in such distinct populations.

> **RECOMMENDATION 4-3: The adequacy of sampling methods aimed at key racial and ethnic groups, as well as the quality of survey mea-surement obtained from them, should be carefully studied and short-comings, where found, remedied for all major national DHHS surveys.**

The DHHS Policy for Improving Race and Ethnicity Data (the "Inclu-sion Policy") clearly states the goal of collecting data on race and ethnicity for all department programs, record collections, and surveys. The depart-ment's household surveys all collect racial and ethnic data in accordance with the Office of Management and Budget (OMB) Standards for Main-taining, Collecting and Presenting Federal Data on Race and Ethnicity (the OMB standards). However, the department's health data frequently come from record systems—either those used to administer a DHHS program (e.g., Medicare) or medical records from clinics, providers, and laborato-ries. The data on race and ethnicity collected through these records are incomplete, inconsistent, and unstandardized. Not all records collect data conforming to the OMB standards for race and ethnicity, and some do not contain such data at all. As a result, the department should enforce the Inclusion Policy and require those programs funded by DHHS that do not currently report data on race and ethnicity to collect such data and to do so in accordance with the OMB standards.

> **RECOMMENDATION 4-4: DHHS should require the inclusion of race and ethnicity in its data systems in accordance with its Policy for Improving Race and Ethnicity Data.**

Data on SEP are needed both to better understand racial and ethnic disparities and to identify and understand health or health care disparities for deprived groups that are not defined by race or ethnicity but that nonetheless experience such disparities. DHHS data systems do not consis-tently collect data on SEP. While the national household surveys generally provide adequate SEP data, obtaining some measures of employment, edu-cation, and insurance coverage, although with limited detail, wealth data are rarely collected. Administrative and medical record systems include very little SEP data; in most cases, only insurance coverage status or method of payment is recorded. Employment status, occupation, and educational attainment are collected, as well. Because of the reporting burden, only limited data can be collected in these systems beyond what is essential.

Nonetheless, the department should consider ways to collect more SEP data in these record systems.

Limited knowledge about health practices and the U.S. health care system or limited communication skills in English are obstacles to obtaining care and understanding diagnoses and treatments. Little information on language use and acculturation (or proxies of it) is collected in national health surveys, although items on these topics could be added. Even less is collected in DHHS administrative records, surveillance systems, and national surveys, although more extensive collection could be justified to facilitate the provision of medical services and information.

RECOMMENDATION 4-5: DHHS should routinely collect measures of socioeconomic position and, where feasible, measures of acculturation and language use.

Weaknesses in a single source of data can often be remedied by linking data from several sources, without the burden of new data collection. For example, by matching SSA earnings records to Medicare claims data, we can study relationships between SEP and health care. Matching might be difficult or inaccurate if common identifiers are not of high quality or are missing. Confidentiality concerns also arise with data linkages because a common identifier is needed in both data sets to link the data, which may increase the possibility that an individual's identity can be recognized. Programs that have linked data have proven that personal identity can successfully be protected with the proper precautions. Where possible, the department should promote relevant data linkages across DHHS agencies and across other agencies or institutions.

RECOMMENDATION 4-6: DHHS should develop a culture of sharing data both within the department and with other federal agencies, toward understanding and reducing disparities in health and health care.

Data on Medicare enrollees, contained in the Enrollment Database (EDB), are crucially important for understanding disparities in health and health care treatment as Medicare expenditures account for about 18 percent of U.S. health care spending (Centers for Medicare and Medicaid Services, 2003). Because of the importance of Medicare data, the panel believes it is crucial to improve Medicare data on race and ethnicity, through initiatives of the Centers for Medicare and Medicaid Services (CMS). For new enrollees, data on race and ethnicity, SEP, and a proxy of acculturation such as language use could be most easily collected at the time of enrollment. This information should also be collected for current enrollees. A very brief questionnaire could be used for both of these efforts. To keep the questionnaire short, it is probably not feasible to collect income and wealth information. However, educational level could be included. A question about language use might also be considered.

To obtain more detailed SEP data for use in analysis, CMS should also obtain records of an enrollee's earnings and employment histories through the wage history files of the SSA. These data show only individual earnings and only for the time period the person worked. They do not necessarily reflect the lifetime earnings of that individual, nor earnings and income available to that person through, for example, the earnings of a spouse. Privacy and confidentiality concerns should always be considered carefully. The panel believes that despite the potential barriers, the CMS and SSA should cooperate to link these two important data sets.

RECOMMENDATION 4-7: The Centers for Medicare and Medicaid Services should develop a program to collect racial, ethnic, and socio-economic position data at the time of enrollment and for current enrollees in the Medicare program.

RECOMMENDATION 4-8: The Centers for Medicare and Medicaid Services should seek from the Social Security Administration a summary of wage data on individuals enrolled in Medicare.

Leadership for Implementing OMB Standards in DHHS Data Systems

The panel found considerable confusion among some groups of data collectors and users regarding the OMB standards for collection of data on race and ethnicity (National Research Council, 2003). To remedy this, DHHS should inform all its agencies, state health agencies, and private entities that collect data for DHHS programs about these new standards. The OMB has published materials to guide the use of the standards and on bridging to old categories of race and ethnicity. DHHS should increase awareness of the OMB standards by disseminating the appropriate OMB materials to the various state and private entities from which data are obtained and assume responsibility for ensuring that the standards are properly and consistently applied throughout the department's data collection systems. DHHS should also develop implementation guidelines specifically aimed at the collection of racial and ethnic data in state and privately based record collection systems.

RECOMMENDATION 4-9: DHHS should prepare and disseminate implementation guidelines for the Office of Management and Budget standards for collecting racial and ethnic data.

Reporting Racial and Ethnic Health Disparities Data in Conjunction with SEP Data

The interrelationship between health and health care and SEP implies that it is important to consider racial and ethnic differences in health and health care within different social and economic backgrounds. Where possible, the panel urges DHHS to report statistics on disparities in health and health care by different levels of SEP.

> **RECOMMENDATION 4-10: DHHS should, in its reports on health and health care, tabulate data on race and ethnicity classified across different levels of socioeconomic position (SEP).**

STATE DATA COLLECTION SYSTEMS

States and U.S. territories are responsible for maintaining numerous health-related data collection systems, including those for vital statistics information (birth and death records); hospital discharge abstracts, which detail information on hospital patients and the diagnoses and treatments they receive; registries, such as the cancer registry system, which provides information on cancer cases and their treatment; and programs such as Medicaid and the State Children's Health Insurance Program (SCHIP). Some states also conduct their own surveys of their populations or have data collection systems for separate programs that provide health insurance and health care. The data in many state-based systems are shared with DHHS for department use in monitoring the health of the nation and administering and evaluating federal programs.

The collection of data on race and ethnicity in these state-based systems is uneven and unstandardized. While the Medicaid and vital records data collection systems follow the OMB minimum standards for racial and ethnic data collection, hospital discharge abstract systems do not; indeed, some do not collect such data at all. Since most of these systems are based on health records, very little information on socioeconomic position or language is collected. With the exception of information on parental education and country of origin for birth records, occupation on death records, and some income data in the Medicaid and SCHIP administrative records, no SEP or language data are routinely collected by states. This is a serious weakness in state-based data collections.

The panel encourages states to require standard racial and ethnic data collection in their health data collection systems, but in a manner that provides states the flexibility to serve their own specific information needs. The OMB standards would allow states the flexibility to collect more detailed information on race and ethnicity. These standards for reporting broad categories of race and ethnicity could then be used by each state to report data to the national level in a uniform manner. The federal government depends heavily on state-based data and, therefore, should provide leadership to states to develop and utilize standards in state data collection systems.

> **RECOMMENDATION 5-1: States should require, at a minimum, the collection of data on race, ethnicity, socioeconomic position, and, where feasible, acculturation and language use.**

There are, of course, barriers to imposing data collection requirements on states. The costs involved in changing reporting and computer systems are not insignificant. Furthermore, racial and ethnic data are often recorded by health care or program administration personnel who are not trained in interviewing (e.g., medical records clerks, providers and health care workers, or funeral directors). Similarly, many data collection systems have incomplete information because respondents may refuse to answer questions about race, ethnicity, acculturation and language use, or SEP or because recorders fail to request or ascertain the data. Many states could use technical assistance in handling missing data, e.g., through statistical imputation or by linking with other data. States also need guidance in implementing the new OMB standards for racial and ethnic data collection, including the bridging of new categories to old categories and the conversion of multiple-category responses (an individual reports he or she has multiracial ancestry) to single-category responses (a single racial ancestry).

Much work on these technical issues has already been conducted by federal statistical agencies. DHHS should use this work to develop guidance for states on how to address these training and methodological issues. DHHS should also provide states with guidance and support for training in recording data on race and ethnicity.

> **RECOMMENDATION 5-2: DHHS should provide guidance and technical assistance to states for the collection and use of data on race, ethnicity, socioeconomic position, and acculturation and language use.**

PRIVATE-SECTOR DATA COLLECTION SYSTEMS

The panel's review of current practices by private health care providers and insurance companies—hospitals, medical group practices, and health insurance plans—revealed that the collection of data on race, ethnicity, language and acculturation, and SEP in the private sector is not common and that, when such information is collected, it is unstandardized. Many hospitals collect racial and ethnic data on patients, and this reporting is fairly complete. However, the data are not reported to state and federal programs in a standardized format and their accuracy for racial and ethnic groups other than white and black is suspect. Some health plans include questions about race, ethnicity, and language use on their enrollment forms, but this information is provided voluntarily by applicants and is often incomplete or missing. Even less is known about what data medical groups collect. Collection of SEP data is just as rare in these privately based data collections. Most often, the only SEP information collected by hospitals is the patient's source of payment. Some health plans ask for level of education on their enrollment forms, but this is not common practice.

Data collected by these private-sector groups could be invaluable for monitoring and better understanding disparities in health and health care. Health insurance plans could use the data to inform quality improvement efforts, to target health promotion and preventive health measures to specific demographic subgroups, and to aid in disease management strategies. Hospitals could also use the data for quality improvement measures or community assessment initiatives.

The failure to collect these data represents an important missed opportunity. The panel believes that intervention from DHHS is needed to ensure that these data are collected. Health plans and hospitals that are interested in collecting these data are concerned that without a federal mandate, such collection will be perceived with suspicion by those who are asked to provide the information. As a result, the collection of these sensitive data items is intermittent and may be suspect in quality. Federal leadership is needed to help legitimize and standardize the collection of these data and could be effected through a DHHS requirement for the reporting of racial, ethnic, SEP, and language data.

RECOMMENDATION 6-1: DHHS should require health insurers, hospitals, and private medical groups to collect data on race, ethnicity, socioeconomic position, and acculturation and language.

DHHS should work with hospitals and health plans to determine the best way to collect data in a standardized way. The coordination of data collection could be complicated because these are private entities. Rule-making under the Health Insurance Portability and Accountability Act (HIPAA) could contribute some uniformity. HIPAA does not currently require the collection of racial and ethnic data, and indeed its strong privacy measures may inhibit such collection. The act does, however, enforce standards on the collection of other data from health services. Thus, for example, a logical starting point for mandating the collection of data on race and ethnicity could be through HIPAA regulations that apply to electronic transactions in two DHHS programs—Medicare and Medicaid. Changes to the required data collection standards are possible through the Designated Standards Maintenance Organizations (DSMOs), the Data Content Committees (DCCs), and other stakeholder organizations. The secretary of DHHS may propose changes to the standards that the DSMOs, DCCs, and other organizations may then consider. The panel suggests further exploration of this avenue for the federal mandating of racial and ethnic data collection.

Whatever standards are chosen should use the OMB standards for their base, supplemented with further detail as needed. DHHS should also work with hospital and health plan-related groups to determine which SEP data are feasible to collect on enrollment or admissions forms. Collection of these data will necessarily be limited as the collection of detailed wealth and income information may impose a burden on providers and on the individuals providing the data. However, an individual's education level may be the easiest and least sensitive item to collect.

In developing standards for data collection, it is critically important to provide clear information about how the data will be used so that individuals providing the data are fully informed. DHHS should work with industry agents and legal experts to develop a list of these uses for hospitals, health plans, and medical groups to give to individuals from whom the data are collected.

RECOMMENDATION 6-2: DHHS should provide leadership in developing standards for collecting data on race, ethnicity, socioeconomic position, and acculturation and language use by health insurers, hospitals, and private medical groups.

Implementation of this report's recommendations would greatly enhance the data infrastructure available for understanding and eliminating disparities. However, if these recommendations cannot be implemented such that high-quality data are produced, linking aggregate-level data on race, ethnicity, SEP, and acculturation and language use may be needed to bridge the gaps. These data aggregated at the level of census geographical units (Zip Code tabulation areas, tracts, or block groups) could be used to proxy individual-level data by linking them to the individual level data that are available.

Suitable confidentiality protections are critical for the use of such linked geocoded data. The precise combination of values of the sociodemographic variables might identify a subject's geographical area and thus pose a risk of disclosure of confidential information about health plan members. Methods have been developed for masking such data by rounding or adding random noise. Such masked data sets can then be analyzed with appropriate corrections for the effects of masking. But development of the specific procedures and parameters required to implement data masking requires particular statistical expertise that is not likely to be found within health insurers.

DHHS could greatly facilitate the routine generation of high-quality, uniform, and nondisclosing geographically linked data sets by providing a linking service that could be used by private- and public-sector health care organizations. Such a service could be administered, for example, through a Web site. The organization would anonymously submit a file containing member addresses, and receive in return a file of masked geographical variables at several levels.

The greatest expertise within the federal government for solving the problems involved in establishing such a service resides in the Bureau of the Census. Within DHHS, the National Center for Health Statistics (NCHS) has been a leader in dealing with confidentiality issues. Alternatively, a private-sector vendor with the necessary geocoding expertise could be recruited, although such vendors do not typically deal with the related confidentiality issues.

RECOMMENDATION 6-3: DHHS should establish a service that would geocode and link addresses of patients or health plan members to census data, with suitable protections of privacy, and make this service available to facilitate development of geographically linked analytic data sets.

1

Introduction

Disparities in health outcomes and health care among different racial and ethnic groups have been well documented in the literature. Infant mortality rates, for example, are higher among African Americans than they are among white Americans (14.1 per 1,000 births for African Americans versus 5.7 per 1,000 for whites; see National Center for Health Statistics, 2002) and the rate of death caused by cardiovascular disease was 146.9 per 100,000 for whites and 230.5 per 100,000 for blacks (U.S. Department of Health and Human Services [U.S. DHHS], 1990). There are substantial differences across racial groups in the percentage of women over the age of 40 who received a mammogram within the past 2 years: according to data from the 2000 National Health Interview Survey, among women at least 40 years old, 71.4 percent of white women received a mammogram compared with 67.8 percent of African American women, 47.3 percent of American Indian or Alaska Native women, and 53.3 percent of Asian women (National Center for Health Statistics, 2003). Disparities among racial and ethnic groups also show up in many chronic and acute disease conditions (e.g., diabetes, cardiovascular disease, obesity, hypertension, and asthma), as well as in the risk factors associated with them. Disparities exist *within* broad racial and ethnic categories as well. For example, among Asian Americans, 34 percent of Koreans have no health insurance, compared with 27 percent of Southeast Asians (Vietnamese, Cambodians, Laotians), 22 percent of South Asians (Indians, Pakistanis, and Bangladeshis), 20 percent of Filipinos, 19 percent of Chinese, and 13 percent of Japanese (Henry J. Kaiser Family Foundation, 2000).

The reasons for these disparities are not fully understood. Differences in access to health care (that is, health insurance coverage and/or the ability to pay for health care) might explain part of these differences. Among persons under 65 years of age in 2001, 19.3 percent of African Americans and 34.8 percent of Hispanics did not have health insurance coverage, whereas 14.7 percent of whites and 17.1 percent of Asians did not (National Center for Health Statistics, 2003). Some racial and ethnic groups tend to have lower income and wealth levels than others and thus may be less able to afford health care. Significant and persistent differences in wealth across racial groups have also been documented (Barsky et al., 2002; Oliver and Shapiro, 1995).

But differences in access are only part of the story. Differences in health care treatment even among the insured and beyond differences in access to health care also contribute to disparities. For example, among Medicare beneficiaries aged 65 and older, white women were more likely to get mammograms and to receive angioplasties than black women (Gornick et al., 1996).

Disparities in health and health care illuminate weaknesses in the health care and public health systems. Interest in better understanding the causes of these differences and formulating strategies to ensure the highest quality of care for everyone has generated significant attention to disparities across racial and ethnic groups. In addition to studies being conducted across many disciplines in a variety of academic, clinical, governmental, and other settings, Congress has initiated several important projects to provide better information on health disparities. In 1999, Congress asked the Institute of Medicine (IOM) to assess the extent of racial and ethnic differences in health care beyond those that are attributable to access to care, to evaluate potential sources of racial and ethnic disparities in health care, and to recommend interventions to eliminate the biases. This study resulted in the 2003 IOM report *Unequal Treatment: Confronting Racial and Ethnic Disparities in Healthcare* (IOM, 2003a). A key finding in this report was that racial and ethnic disparities exist beyond what can be attributed to differences in access to care. The panel also found that these disparities contribute to worse outcomes in many cases (IOM, 2003a). This study also examined the factors that contribute to disparities and offered guidance on interventions to reduce and eliminate disparities.

In 1999, Congress required the Agency for Healthcare Research and Quality to produce an annual National Healthcare Disparities Report (NHDR) that would monitor disparities in health care by race, ethnicity, socioeconomic status, and geography. Congress also asked the IOM to provide guidance on the development of the NHDR, a study that culminated in the 2002 production of *Guidance for the National Healthcare Disparities Report* (IOM, 2002). The first of the NHDR annual reports was released in December 2003 (U.S.

DHHS, 2003a).[1] The purpose of this series of reports is to "track prevailing disparities in health care delivery as they relate to racial factors and socioeconomic factors in priority populations" (P.L. 106-129).

DHHS has implemented initiatives aimed at better understanding and addressing disparities. Eliminating health disparities is one of two primary goals of the Healthy People 2010 program. Some agencies within DHHS have implemented initiatives of their own to correspond to Healthy People 2010. The department has also initiated an educational campaign called Closing the Health Gap. A key element of this campaign is Take a Loved One to the Doctor Day, which is an effort to "encourage individuals to take charge of their health by visiting a health professional (a doctor, a nurse, a nurse practitioner, a physician assistant, or another health provider), making an appointment for a visit, attending a health event in the community, or helping a friend, neighbor, or family member do the same" (http://www.healthgap.omhrc.gov/index.htm).

DATA TO SUPPORT HEALTH DISPARITIES INTERVENTIONS AND RESEARCH

The availability of high-quality data on race, ethnicity, and other characteristics of individuals receiving health care is critical to documenting disparities in health and health care. But there are many weaknesses in the data sources currently available.

National-level surveys sponsored or conducted by the federal government are rich in information on health and health care outcomes as well as other characteristics of individuals; but while most have large enough sample sizes to obtain reliable information about broad racial and ethnic groups (i.e., blacks and whites), sample sizes are often not large enough for analyses of smaller racial groups (e.g., American Indian and Alaska Native) or for analyses within some of the racial and ethnic groups (e.g., to analyze differences between individuals of Mexican descent and individuals from other Hispanic backgrounds).

Data from Medicare claims and enrollment files have been widely used for analysis of racial and ethnic disparities, but such data are not available for all enrollees or potential enrollees, nor do they include information to permit analysis for more refined categorizations of race and ethnicity.

---

[1]We note that two versions of this report were released; the executive summaries of these versions were the only differences in the two releases. The first version was released in December 2003; a second version was released in February 2004. The second version contained what was the original executive summary of the report, which had been changed in the process of departmental review (see Pear, 2004). The key findings of these two releases of the report are different (see http://www.house.gov/reform/min/politicsandscience/example_disparities.htm for both versions of the executive summary).

While state-based sources of data—vital records, administrative data from Medicaid, the State Children's Health Insurance Program (SCHIP), and disease registry systems, for example—are potentially valuable for the analysis of health disparities, they often do not contain standardized data on race and ethnicity and contain very little information on other characteristics of individuals that would be used in an analysis. As a result, data from these sources often need to be matched with data from other sources.

Finally, data on health and health care are obtained from private records (those of health insurance plans, hospitals, and medical groups). But the racial and ethnic data on hospital records are unstandardized and include little information on the economic and social standing of patients. Health plans collect very little data on race and ethnicity, although there is some movement among plans to begin to collect such data.

PANEL CHARGE

Concerns about the data infrastructure for analyzing, understanding, and eliminating racial and ethnic disparities motivated Congress, in the Minority Health and Health Disparities Research and Education Act of 2000, to ask the National Academies to conduct a comprehensive study of the adequacy of Department of Health and Human Services (DHHS) data collection systems for measuring racial, ethnic, and socioeconomic disparities in health. In response to this request, the Office of the Assistant Secretary for Planning and Evaluation (ASPE) asked the Committee on National Statistics (CNSTAT) of the National Academies to convene a panel of experts to review DHHS data systems. The panel was charged to review the collection of data on race and ethnicity in data collection or reporting systems of the programs or activities of DHHS. These include other federal data collection systems (such as that of the Social Security Administration) with which the department interacts to collect data on race and ethnicity, as well as such systems in states and in the private health care sector (U.S. DHHS, 2003b). The panel was asked to:

• identify the data needed both to support efforts to evaluate the effects of race, ethnicity, and socioeconomic position on access to health care and on disparities in health and to enforce existing protections for equal access to heath care;
• assess the effectiveness of the data systems and collection practices of DHHS and the effectiveness of selected systems and practices of other federal, state, and local agencies and the private sector in collecting and analyzing such data; and
• identify critical gaps in the data and suggest ways in which they could be filled, including the possible establishment of new systems.

ASPE and CNSTAT developed the charge for the study based on this legislation and on the department's own needs for review of its data systems, giving the panel the flexibility to review related data needs as they arose.

We note some specific distinctions the panel made in interpreting its charge. First, the panel reviewed a very broad set of data collection systems both within and outside DHHS. These systems include health surveys, administrative records from programs operated by DHHS and the states (e.g., Medicare, Medicaid, and SCHIP), and records from private data systems such as health insurance records, hospital discharge abstracts, and physician and medical group records. The research purposes and uses of these data collection systems are quite varied—some are used to understand broad determinants of health (e.g., the effect of income on mortality) while others are used to understand very specific outcomes of health care treatment (e.g., the effects of ethnicity and race on medical outcomes of patients with hypertension or diabetes). The panel focused on the collection of data on race, ethnicity, and socioeconomic position, as called for in the originating legislation and charge to the panel. The panel also considered the collection of data on acculturation and language use because it believed these to be important covariates in understanding racial, ethnic, and socioeconomic aspects of health and health care. To comprehensively study health and health care, much more data are needed on individuals (e.g., genetic, behavioral, environmental, and cultural), and their health care treatments (e.g., what treatment they received, the cultural competency of the health care professionals administering the treatment, etc.). Collection of these data is affected by issues of race, ethnicity, and socioeconomic position because concepts of health and health care may differ or may be communicated in different ways in different cultural and socioeconomic groups in the United States. For example, the Western medical concepts of mental and physical health may not be consonant with all survey respondents, or specific concepts such as "family planning" or "gestational age" may be interpreted quite differently by Hispanic and Anglo respondents. A review of these issues is, however, beyond the scope of this panel, whose charge was to review the collection of basic racial, ethnic, and socioeconomic characteristics rather than to explore fully their implications for health research.

PREVIOUS REVIEWS OF DATA NEEDS

The panel's work comes on the heels of previous DHHS efforts to examine collection of racial and ethnic data.[2] In 1999, the department

---

[2]We also note that the Healthcare Equality and Accountability Act, which has been introduced in Congress, calls for DHHS to require the collection of data on race, ethnicity, and primary language of applicants to DHHS health-related programs.

implemented the DHHS Policy for Improving Race and Ethnicity Data, or the "Inclusion Policy," which requires the inclusion of information on race and ethnicity in all DHHS-funded and -sponsored data collection systems. Also in 1999, the DHHS Data Council and the Data Work Group of the DHHS Initiative to Eliminate Racial and Ethnic Disparities in Health produced a report called *Improving the Collection and Use of Racial and Ethnic Data in Health and Human Services* (U.S. DHHS, 1999). This report contained recommendations for DHHS to improve the collection, analysis and interpretation, dissemination and use, research and maintenance of racial and ethnic data. Previous reports produced by DHHS or their contractors have also addressed the topic of data collection for measuring racial and ethnic disparities (Waksberg, Levine, and Marker, 2000; U.S. Public Health Service, 1992, 1993; U.S. DHHS, 1985). The National Committee on Vital and Health Statistics, an advisory committee to DHHS, has devoted serious attention to the collection of data on race and ethnicity within DHHS by holding hearings around the nation, writing reports, and making recommendations that encourage the collection of such data.

The panel considered these DHHS reports during its information-gathering phase and also invited experts on various data-related topics—including the Health Insurance Portability and Accountability Act (HIPAA) and the National Disparities Report Card—as well as members of the National Committee on Vital and Health Statistics Subgroup on Populations to brief the panel in open meetings. The panel also sponsored a 2-day Workshop on Improving Racial and Ethnic Data in Health, which focused on state-based and private sector-based data collections and resulted in a summary report (NRC, 2003). The agenda for this workshop is included in Appendix B. The panel commissioned a paper for this report to review the collection of socioeconomic position measures in health and health care databases. This paper, Recommendations on the Use of Socioeconomic Position Indicators to Better Understand Racial Inequalities in Health (O'Campo and Burke), is included in Appendix C. Four papers also were commissioned for the workshop: The Role of Racial and Ethnic Data Collection in Eliminating Disparities in Health Care (Fremont and Lurie); State Collection of Racial and Ethnic Data (Geppert, Singer, Buechner, Ranbom, Suarez, and Xu); Collection of Data on Race and Ethnicity by Private-Sector Organizations: Hospitals, Health Plans, and Medical Groups (Nerenz and Currier); and Racial and Ethnic Data Collection by Health Plans (Bocchino). These papers are included in Appendixes D-G of this report.

THE REPORT

The rest of this report contains the panel's assessment of the strengths and weaknesses of racial and ethnic data collection efforts in the health and health care fields. The panel believes very strongly that social and economic factors as well as language and acculturation factors play an important role in understanding racial and ethnic disparities in health and health care, and therefore considers the collection of measures of these concepts just as important as measures of race and ethnicity. Chapter 2 sets the stage by briefly reviewing some of the literature on such disparities, citing examples of disparities, and making the case for the importance of collecting data to monitor and understand them and to develop interventions to eliminate them. Chapter 3 reviews measures of race and ethnicity and recent regulations governing the collection of these data. The chapter also briefly reviews measures of socioeconomic position, language, and acculturation that are used to understand the relationship between health and these constructs.

The remaining chapters of the report review current practices in the collection of data on race, ethnicity, socioeconomic position (SEP), and language and acculturation in the health care field, highlight weaknesses, and recommend improvements. The discussion is organized by the source of the data—federal, state, and private-sector sources. There is occasional overlap in this partition, as some data generated by private-sector entities are collected and organized by states, sometimes with federal funding support. Recommendations for improved data collection are given in each of the last three chapters. In making these recommendations, the panel did not consider specific assessments of the cost of improved data collection but did broadly consider the costs of data collection among different types of data collection systems.

The primary focus of the panel was on data collection systems within DHHS, over which the department has the most control. Chapter 4 reviews data collected by DHHS. The panel also reviewed data collected by states and by private sector entities. Both state health departments and private sector entities have roles to play in the collection of data for research on disparities. States are responsible for the collection and maintenance of a number of data systems in health and health care—many of which are then forwarded to DHHS for use on a national level (e.g., vital records and hospital discharge abstracts). Chapter 5 reviews data collected by state governments. Private entities also collect data relevant for researching disparities. These records of diagnoses, treatments, and insurance enrollments and claims are especially important sources of information on health care. Some of these records are the basis of federal and state-based data collection systems. Although outside of DHHS and state health agencies, these privately based records are an important component of data systems needed for the federal government and for state governments to monitor the health of their respective populations. Chapter 6 reviews data collected by private sector organizations.

2

The Importance of Data on Race, Ethnicity, Socioeconomic Position, and Acculturation in Understanding Disparities in Health and Health Care

Serious disparities in health and in access and utilization of health care and medical treatment have been found across racial and ethnic groups and economic and social strata in the United States. Disparities linked to language proficiency and acculturation have also been found. As a precursor to our discussion of data needs for understanding these differences, we first consider why social and economic characteristics such as race, ethnicity, socioeconomic position (SEP), and language use and acculturation are important for understanding disparities in health and health care. Throughout the report, we use the term SEP (instead of the widely used alternative term socioeconomic status, or SES) to encompass a broad set of socioeconomic characteristics including income, wealth, and education. Although often used loosely as synonymous with SEP as defined here, SES is sometimes thought to refer solely to the narrower concept of status, which has connotations of a specific standing in society.

We begin by reviewing the literature on disparities. Next, we discus the panel's definition of disparities. The key dimensions the panel was charged to consider—race, ethnicity, and socioeconomic position—as well as language use and acculturation are then discussed. The chapter concludes by highlighting the importance of better understanding the causes and consequences of disparities in health and health care.

DISPARITIES IN HEALTH AND HEALTH CARE

Examples of Racial, Ethnic, and Socioeconomic Position Disparities

An extensive body of literature covering a number of health and social science disciplines has documented persistent racial, ethnic, and socioeconomic disparities in health status and health care in the United States. For some measures of health and health care, these disparities have existed over a long period of time, or at least since data were available to measure them; in some cases they have decreased over time, and in others increased.

The causes of these disparities are not well understood. Differences in economic conditions across racial and ethnic groups (in general, racial and ethnic minorities and recent immigrants are poorer than nonminorities) probably contribute to disparities, as they are likely to result in less access to health care, inability to afford higher-quality care, and greater exposure to harmful occupational and environmental factors. Differences in education may contribute to disparities, as may health-related behavior patterns (e.g., diet, exercise). And, of course, bias and discrimination may also contribute to racial and ethnic disparities. In this section, we highlight a few examples of disparity problems.[1]

Table 2-1 shows infant mortality rates by racial and Hispanic origin of the infant's mother from 1983-2000. In the most recent period shown, 1998-2000, non-Hispanic black infants had the highest infant mortality rates by far, with nearly 14 deaths per thousand live births. This contrasts sharply with the infant mortality rate among non-Hispanic whites, which is just under 6 deaths per thousand live births. The table also shows wide variation within broad racial and ethnic groups. For example, the overall infant mortality rate for Asian and Pacific Islanders is 5 deaths per thousand live births. Within this category, however, the infant mortality rate for Chinese Americans is 3.5 deaths per thousand live births, while the infant mortality rate for Filipinos is almost 6 deaths per thousand live births and the infant mortality rate for Hawaiians is almost 9 (National Center for Health Statistics, 2003, p. 122). With other health measures, the combined categorization of Asians and Pacific Islanders into a single subgroup has also masked variation among ethnicities within this subgroup—for example, Pacific Islanders have elevated levels of morbidity and mortality compared to the U.S. population (Frisbie, Cho, and Hummer, 2001).

---

[1]There are also substantial urban versus rural disparities in health and health care (Ricketts, 2002; Skinner et al., 2003). These disparities are also of concern to the federal government and were discussed in the recently released National Healthcare Disparities Report (U.S. DHHS, 2003a). The data collection needs for understanding geographic disparities are beyond the scope of this panel's charge, but better measurement of racial, ethnic, and socioeconomic disparities should help in the measurement and interpretation of geographical disparities.

TABLE 2-1 Infant Mortality Rates by Race and Hispanic Origin

Race and Hispanic Origin of Mother	Infant Deaths per 1,000 Live Births				
	1983-1985	1986-1988	1989-1991	1995-1997	1998-2000
All mothers	10.6	9.8	9.0	7.4	7.0
White	9.0	8.2	7.4	6.1	5.8
Black or African American	18.7	17.9	17.1	14.1	13.8
American Indian or Alaska Native	13.9	13.2	12.6	9.2	9.0
Asian or Pacific Islander	8.3	7.3	6.6	5.1	5.1
Chinese	7.4	5.8	5.1	3.3	3.5
Japanese	6.0	6.9	5.3	4.9	3.8
Filipino	8.2	6.9	6.4	5.7	5.9
Hawaiian	11.3	11.1	9.0	7.0	8.7
Other Asian or Pacific Islander	8.6	7.6	7.0	5.4	5.2
Hispanic or Latino	9.2	8.3	7.5	6.1	5.7
Mexican	8.8	7.9	7.2	5.9	5.5
Puerto Rican	12.3	11.1	10.4	8.5	8.1
Cuban	8.0	7.3	6.2	5.3	4.3
Central and South American	8.2	7.5	6.6	5.3	4.9
Other and unknown Hispanic or Latino	9.8	9.0	8.2	7.1	6.9
Not Hispanic or Latino					
White	8.8	8.1	7.3	6.1	5.9
Black or African American	18.5	17.9	17.2	14.2	13.9

SOURCE: National Center for Health Statistics (2003).

Racial and ethnic differences in incidence and death rates for different diseases also exist. Box 2-1 shows prevalence and death rates for diabetes across racial and ethnic groups. Rates of complications from Type 2 diabetes mellitus differ across racial and ethnic groups, even among members of the same HMO and after controlling for differences in income, education, health behavior, and clinical characteristics (Karter et al., 2002).

Disparities in access to and utilization of health care services between racial and ethnic groups may well contribute to disparities. Minorities, especially Hispanics, are much more likely than whites to lack any type of health insurance coverage. Among those under the age of 65 in 2001, 34.8 percent of Hispanics and Latinos had no health insurance coverage (National Center for Health Statistics, 2003, Table 129), compared with 11.9 percent of non-Hispanic whites and 19.2 percent of non-Hispanic blacks.

But utilization of health care services differs across races even among those who are insured. For example, Table 2-2, shows differences in the rates at which black and white Medicare enrollees receive selected services and shows that black Medicare enrollees are less likely to receive preventive care. Blacks in general are less likely than whites to visit a physician's office, to see an ophthalmologist, and to have a sigmoidoscopy or colonoscopy. As a result, as Gornick (2002) points out, blacks are more likely to end up having surgery for complications of poorly controlled chronic illnesses—for example, amputations of limbs for diabetes or treatment for retinal lesions. Schneider, Zaslavsky, and Epstein (2002) also found that among Medicare

---

BOX 2-1
Prevalence of Diabetes Among Adults Aged 20 and Older
(percent of population)

White	7.4
Hispanic	13.6
Black	15.0
American Indian/Alaska Native	18.8

Age-Adjusted Death Rates
per 100,000 People for Diabetes Mellitus

Asian/Pacific Islander	16.4
White	22.8
Hispanic	36.9
American Indian/Alaska Native	41.5
Black	49.5

SOURCE: National Center for Health Statistics (2003).

TABLE 2-2 Ratio of Percent of Black Women Receiving Mammogram, Flu Shot, and Pap Smear 65 Years of Age and Over, 1998

	Mammogram	Flu Shot	Pap Smear
Black to White Ratio of % Receiving Service			
Unadjusted	0.92	0.74	0.88
Adjusted for Income	0.97	0.77	0.92
Adjusted for Education	0.98	0.75	0.99

SOURCE: Unpublished tabulations from the 1998 Medicare Current Beneficiary Survey (Gornick, 2002).

beneficiaries in managed care health plans, black patients were less likely to receive breast cancer screening than white patients, even after adjusting for SES. One study has found that of patients with end-stage renal disease, black Americans have lower rates of referral for renal transplantation than white Americans (Epstein et al., 2000). Another study showed that black Americans are referred for cardiac catheterization less frequently than white Americans (Peterson et al., 1994).

Disparities in health outcomes also exist across levels of economic and social position. Several investigators have found that higher incomes are associated with lower mortality (Sorlie, Backlund, and Keller, 1995; Deaton and Paxson, 2001) and that people with higher incomes can expect to live longer than those with lower incomes (National Institutes of Health, 1992). Some research indicates a correlation between SEP and cancer incidence and mortality (Singh et al., 2003): those from high-poverty areas had higher incidence of cancer, later-stage diagnosis, and higher mortality rates than those from low-poverty areas. But the relationship does not hold for all types of cancers. For example, incidence and mortality from melanoma are higher in low-poverty areas, probably reflecting higher levels of income and wealth among whites, who are at greater risk for melanoma (Singh et al., 2003). This study also found that women in higher poverty areas were less likely to be diagnosed for breast cancer than women in low poverty areas, but had higher mortality rates from breast cancer.

There is also some evidence that prolonged experience of lower economic status, or at certain stages of the life course, translates into worse health outcomes later in life (Marmot and Wadsworth, 1997). Case, Lubotsky, and Paxson (2002) show that the relationship between low SEP and poor health is stronger for older children than for younger children. Currie and Stabile (2002) suggest that this result may occur because low-SES children receive a greater number of negative health shocks (e.g., accidents, incidence of disease) during childhood and these shocks accumulate into worse outcomes later in childhood.

Health outcomes and health-related behaviors also differ by levels of education. Figure 2-1 shows the age-adjusted prevalence of cigarette smoking by persons aged 25 and over by education level. While 31.9 percent of high school dropouts smoke, only 10.9 percent of those with college degrees smoke. For this and other reasons, mortality risk diminishes for those who have higher levels of education (Sorlie, Backlund, and Keller, 1995). Occupation has also been found to be associated with mortality differences. For example, in a longitudinal study of British civil servants, those who held professional or executive positions had the lowest mortality rate for coronary heart disease, neoplasms, and nonneoplasms compared with clerical and other occupations (van Rossum et al., 2000).

Race, ethnicity, and socioeconomic position are interrelated in the population, but each makes its independent contribution to health. For example, Table 2-3 shows rates of hypertension and overweight among white, black, and Mexican American women by economic status (poor, near poor, and middle-to-high income). For white women, lower income is associated with a markedly higher prevalence of hypertension and overweight. For black women, lower income is also associated with a markedly higher prevalence of hypertension, but not with the prevalence of overweight. Mexican American women are less likely to suffer from hypertension than either white or black women and there are essentially no income-related differences in hypertension rates among them. Mexican American women are more often overweight than white women but less than black

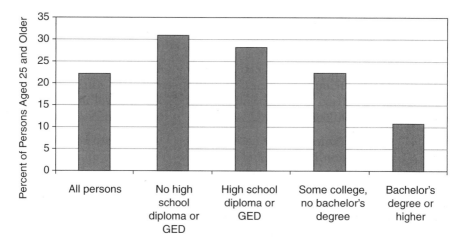

FIGURE 2-1 Age-adjusted cigarette smoking prevalence by education level, 2001.
DATA SOURCE: National Center for Health Statistics (2003).

TABLE 2-3 Age-Adjusted Rates of Hypertension and Overweight, by Race and Ethnicity and Average Annual Income: Women in the United States Aged ≥20 Years, 1988-1994

Income Level	Hypertension, Percent			Overweight, Percent		
	White	Black	Mexican American	White	Black	Mexican American
All (ages 20-74)	19.3	34.2	22.0	32.5	53.3	51.8
Poor	30.2	39.9	24.5	42.0	55.0	54.9
Near poor	23.9	35.9	22.4	36.6	51.0	48.7
Middle/high income	20.2	30.0	25.2	30.0	52.4	45.3

SOURCE: Williams (2002).

women, with significant variation across income levels. Thus, race, ethnicity, and income interact in complex ways to affect these health outcomes.

In Chapter 1, we reported that black women were less likely to receive mammograms than white women. Table 2-2 shows that these racial differences may be partially explained by differences in income and in education, at least for women over the age of 65. The ratio of the percentage of black women to the percentage of white women receiving mammograms was 0.92 in 1998. However, when adjusted for income, this ratio increases to 0.97. Similarly, when education level is controlled, the ratio of black to white rates increases to 0.98.

Adjustments for income and education have little effect on black to white ratios for receiving flu shots (0.74 to 0.77 for income and 0.75 for education). On the other hand, adjusting for education nearly eliminates the differential between blacks and whites for pap smears; controlling for income, however, increases the black to white ratio only slightly.

A growing body of literature has documented disparities across races and ethnicities in health care treatment and quality of treatment (IOM, 2003a). Even after controlling for differences in other patient background characteristics, Schneider, Zaslavsky, and Epstein (2002) found racial differences among Medicare beneficiaries in managed care plans both in their receipt of beta-blockers after a myocardial infarction and in follow-up after hospitalization for mental illness. Black patients were less likely than white patients to receive beta-blockers and follow-up after hospitalization.

Disparities in health outcomes by the level of acculturation and proficiency with the English language have also been clearly documented in the research literature. The link between acculturation and health and health care is observed for many racial and ethnic groups, including Asian and Pacific Islanders and Hispanics (Clark and Hofsess, 1998; Balcazar and Qian, 2000). A variety of studies have demonstrated that high levels of acculturation among Hispanics are associated with higher rates of low birthweight and intrauterine growth retardation, higher rates of adolescent pregnancies, early sexual initiation, and high levels of smoking, drugs, and alcohol consumption in adolescents. On the other hand, low acculturation is related to less access to health care and preventive services, and a lower probability of outpatient care for mental health problems (Clark and Hofsess, 1998). The patterns of low access to health care utilization and low use of preventive health services by less acculturated individuals cut across racial groups including many Asian subgroups such as Vietnamese, Chinese, Korean, and Japanese (Yi, 1995; Yeh, 2003).

Language use has been shown to be a powerful predictor of health care utilization and health among many racial and ethnic groups (Unger et al., 2000; English, Kharrazi, and Guendelman, 1997; Yu et al., 2002; Fiscella et al. 2002; Byrd, Balcazar, and Hummer, 2001). Language proficiency is also a very important factor in relation to enhancing patient-provider relationships and thus affecting health care outcomes. Indeed, a recent study found that in cases where physicians spoke the same language as the patient, the patient reported better physical functioning, psychological well-being, and health perceptions, as well as less pain than in cases where the physician and patient did not speak the same language (Perez-Stable, Napoles-Springer, and Miramontes, 1997).

Initiatives to Monitor and to Address the Problem of Disparities

Several major federal initiatives are aimed at monitoring trends in disparities, understanding the causes of disparities, and planning programs to reduce disparities when they are found. Most of these efforts focus on elimination of disparities across racial and ethnic groups rather than economic and social groups.

Eliminating health disparities across racial, ethnic, and education or income groups (and among other population groups as well) is one of two primary goals of Healthy People 2010, a long-term national agenda aimed at improving health in the United States (see U.S. DHHS, 2003a, and http://www.healthypeople.gov/). One of the initiative's goals is to eliminate health disparities by the year 2010. The goals and objectives were developed by groups of experts directed by the secretary of the Department of Health and Human Services, in partnership with DHHS agencies, various state and territorial groups, and other nonfederal government agencies, with input from relevant business groups and community organizations. The initiative has set many objectives with corresponding targets for specific improvements to be achieved through health and program interventions. Success in meeting these targets will be tracked with indicators of health and health status from a number of different data sources.

Various DHHS agencies also are involved in initiatives corresponding to Healthy People 2010. The Centers for Disease Control and Prevention (CDC) is housing REACH 2010, a two-phased, 5-year departmental program demonstration project to support community coalitions in designing, implementing, and evaluating community-driven strategies to eliminate health disparities. The National Institutes of Health have a Strategic Research Plan to Reduce and Ultimately Eliminate Health Disparities. This 5-year plan for enhancing research, research infrastructure, and public information and community outreach outlines NIH's objectives for reducing and eliminating disparities.

Disparities in health care have been recognized only recently relative to disparities in health. The IOM report (2002) on disparities in health care contained many recommendations for health system interventions, legal and regulatory changes, education promotion, data collection, and future research to begin to eliminate disparities. To date, however, there has been no official response to the report from DHHS.

In 1999, Congress required the Agency for Healthcare Research and Quality of DHHS to develop a National Healthcare Disparities Report (NHDR).[2] The first of these annual reports was released in December 2003. The purpose of the report is to track the extent of disparities in health care and monitor whether progress has been made toward eliminating them.

One major initiative specific to racial and ethnic data collection efforts is the DHHS Inclusion Policy, implemented in 1999 (http://aspe.os.dhhs.gov/datacncl/inclusn.htm). This policy requires the inclusion of information on race and ethnicity in all DHHS-funded and sponsored data collection systems. It was implemented to monitor whether programs are administered in a nondiscriminatory manner and to ensure that standard racial and ethnic data are available to help coordinate responses to health and social service issues. The policy requires that the Office of Management and Budget (OMB) standards for racial and ethnic data collection be used.[3]

In November 2003, the Healthcare Equality and Accountability Act was introduced in the Senate and the House.[4] This act would, among other things, require any health-related program administered, funded, or reimbursed by DHHS to collect data on the race, ethnicity, and primary language of each applicant for and recipient of health-related assistance.

---

[2]See http://www.qualitytools.ahrq.gov/.
[3]These standards, which will be discussed further in the next chapter, were adopted in 1997 and can be found at http://www.whitehouse.gov/omb/fedreg/ombdir15.html.
[4]The Senate version is S.1883. The House version is H.R. 3459.

What Are Disparities?

Throughout this report, we use the term *disparities* to indicate differences in health and health care, where *health* refers to the status of an individual's condition (i.e., the presence of a health condition or illness, such as high blood pressure, asthma, overweight, drug use) and *health care* refers to the process of treating an illness or injury. Some disparities may be inequitable, but not all are. The assessment of whether a disparity in health is inequitable involves societal values as well as scientific explanation. A disparity due to discrimination in preventive care might be regarded as obviously inequitable. A disparity due to differing prevalence of a genetically based disease might not be regarded as inequitable, but if disproportionately few resources were devoted to research and treatment for diseases prevalent in certain groups, an issue of equity might be perceived.

Different perspectives on what constitutes a disparity in health and health care arise from a variety of policy and scientific positions.[5] For example, IOM (2003a) and Gomes and McGuire (2001) define disparities in health care as differences in the treatment of individuals from different groups when these differences are not justified by clinical appropriateness or by patient preference. Using this definition, if members of a group less often receive coronary artery bypass graft because their health status more frequently contraindicates major surgery, this would reflect a health disparity, but not a health care disparity. Similarly, if members of one group more often refuse surgery for a condition while members of another group choose to have the surgery, there is no disparity in health care access.[6] However, according to this definition, a disparity would exist if two patients had all the same medical conditions and would have chosen the same treatment if they were offered the same options, but one patient was not given the treatment while the other was. Although this particular definition of disparities in health care is not universally accepted, the related data variables that are required for studying disparities (as defined by Gomes and McGuire, 2001) would be similar under other definitions.

A disparity in health care may or may not be due to conscious or unconscious discrimination;[7] there may be other causes such as lack of access to particular kinds of health care, poor communication between the patient and the provider, lack of information, individual behavior, and

---

[5]See also Carter-Pokras and Baquet (2002).

[6]Patient preferences may themselves be affected by past experiences of discriminatory behavior in a medical or other setting. For example, if a person feels she has experienced discrimination in the past, she may be reluctant to trust medical professionals in future encounters, and therefore avoids seeking some medical services.

[7]The IOM definition asserts that disparities are caused by two factors—either discrimination or health care system and legal and regulatory factors.

patient preferences that may in part be shaped by past experience of discrimination or deprivation or by cultural, geographical, and other patient background factors.

The panel is charged with reviewing the availability and quality of data used to measure and better understand disparities. It is therefore outside the scope of the panel to make a scientific or policy assessment of whether a specific disparity is inequitable. Instead of adopting a particular definition of disparities, the panel has reviewed the data needed to maintain a monitoring system and to conduct analyses of disparities in health and health care that will allow policy and scientific analysis of the factors that contribute to disparities.

Although disparities in health and in health care are usually discussed together in this report, the panel recognizes that these two types of disparities, while both important, are not equivalent, nor are the data sources required for understanding the two always the same. For example, disparities in health care are routinely studied by examining data from the health care system, while studies of disparities in health also require information on individuals who do not participate in the health care system. Thus, the full range of available health data sources, from administrative data sets to population surveys, is critical to the understanding of disparities and will be discussed in this report.

FOUR KEY DIMENSIONS OF DISPARITIES

The panel was asked to review the adequacy of current policies and practices relating to the collection and availability of data on race, ethnicity, and SEP. We believe that another key dimension to the understanding of disparities is the degree to which an individual is acculturated into U.S. society, including the individual's English language proficiency, and so we have added that dimension to our considerations. Each of these four dimensions is important in itself as well as in relationship to the other dimensions. There are, of course, other factors that determine individual and population health. These include individual characteristics and behaviors and ecological, policy, and health care system factors. There are several models of the determinants of health (reviewed by IOM, 2003b) but we focus on data needs related to race, ethnicity, SEP, and acculturation and language use.

In this section we discuss definitions of race, ethnicity, SEP, and language use and acculturation and their importance for understanding disparities in health and health care. The definitions of these concepts is, of course, the subject of much scholarly literature and media attention. This report does not intend to review that literature completely nor attempt to develop original definitions of these concepts.

Race and Ethnicity

The modern concept of race and the classification of individuals into racial categories has predominantly been based upon phenotypic or observable characterizations, such as skin color or facial characteristics. However, through recent advancements in genetics and biomedicine, as well as a long line of literature in anthropology, sociology, and other behavioral sciences, it has become more commonly recognized that race is not a meaningful biological marker (Lewontin, 1972; American Association of Physical Anthropology, 1996; Cooper and David, 1986).[8] Rather, it is now more commonly recognized that racial classifications are social markers, constructed through a social process (Omi and Winant, 1986; NRC, 2004). The "social constructivist" point of view argues that even though race does not have a meaningful biological definition, it is still an important social and political marker because it reflects, however imperfectly, categories that play important roles in the distribution of power and wealth, discrimination, cultural and personal identity, and group solidarity (American Sociological Association, 2003; Harris, 2002).

Ethnicity usually refers to groupings defined by a common national or regional origin, with a consequent assumed commonality (to some degree) of culture and language. There is overlap in the markers used to indicate race and those used to indicate ethnicity, but race is distinct from ethnicity in that it has an ascribed physical or biological component. For example, *Hispanic* is a general ethnic term for people of Spanish-speaking origin. There are, however, white Hispanics and black Hispanics as well as many different Hispanic ethnicities, such as Puerto Ricans, Mexicans, and Cubans. There are also many ethnicities within categories of non-Hispanic whites, non-Hispanic blacks, and Asians.

Ethnic and racial categorizations are fluid both because the ways they are measured has changed over time and because the context in which they are measured may affect how individuals are classified or classify themselves. For example, Tutsis and Hutus in the United States would be considered black ethnic groups, but in Rwanda would be considered races because in that country, the difference is an important social and political distinction (Harris, 2002). Furthermore, the same individuals may identify themselves as members of different races or ethnic groups in different settings; for example, one person in different circumstances might identify as Taiwanese, Chinese, or Asian American. Recognizing that race and ethnicity are

---

[8]Genetics has produced some evidence that there is systematic variation between groups from different continents of origin (Bamshad and Olson, 2003), although most of the genetic variation is within continental populations (Cooper, Kaufman, and Ward, 2003). There is still active debate, however, over the importance of this variation in a medical setting (see Cooper, Kaufman, and Ward, 2003, and Burchard et al., 2003).

both essentially in the eye of the beholder (whether ego or alter), we often refer to the combined concept of race and ethnicity. In this report, we will use these two together, recognizing that they do not quite have equivalent social meanings.

If race and ethnicity are constructs created to distinguish groups socially but not biologically, why is it so important to understand relationships between race and ethnicity and health outcomes? It is precisely because race and ethnicity are socially and politically constructed categorizations. They have been used to discriminate in the labor and housing markets, in education, and in health care, and they have also been the basis of segregation. For example, as recently as the 1960s, many hospitals in the South were still formally segregated into sections where blacks received care and sections where whites received care (D.B. Smith, 1999). Although no longer legally sanctioned, de facto residential segregation is still widespread in the United States (Massey, 2001) and contributes to segregation in primary and secondary schooling as well as negative employment outcomes, especially for poor inner-city minorities (Wilson, 1987; Massey and Denton, 1993). Segregation in health care facilities is also present in some areas (Fennell, Miller, and Mor, 2000). Finally, there are important links, both historical and current, between race and ethnicity and economic deprivation in the United States. The lasting effects of slavery, discrimination, and segregation have contributed to greater deprivation in some racial groups (e.g., American Indians and blacks). Further, immigrants to this country often come from less well-off backgrounds or their credentials and status in their countries of origin carry less weight (e.g., immigrant doctors who cannot practice in this country without being relicensed).

The recently released report of a National Academies panel charged with defining racial discrimination and assessing methodologies to measure it concluded that because race is a salient aspect of social, political, and economic life in the United States, it is necessary to collect data on race and ethnicity to monitor and understand differences among population groups (NRC, 2004).

Socioeconomic Position (SEP)

The economic and social resources of individuals are strongly related to their health, health care, and access to health care. Studies cited earlier in this chapter have shown that individuals with greater economic and social resources generally have better health and are better equipped (financially and educationally) to navigate the health care system. Thus SEP is an important dimension to health disparities.

SEP is a complex concept, encompassing a number of elements of a person's position in society, including economic resources (earnings, income, and wealth), social resources (social networks and connections to community resources), education (formal credentials, communication skills, and health information), and occupation. Some of these elements may evolve over the life course; for example, income and wealth can change greatly over the course of one's life. Furthermore, the importance of income and wealth may be more important to health and health care at different points over the life course; for example, deprivation of resources may have more negative effects on health at infancy or during childhood than later in life. We use the term SEP throughout the report to refer to the set of these and other related elements of economic and social standing.

SEP affects health in many ways. Socioeconomic deprivation might be marked by increased exposure to hazardous environmental, occupational, or public health conditions, to poorer or more dangerous neighborhoods, or to poor nutrition, with obvious implications for personal health. Furthermore, mechanisms that may be used to cope with deprivation may lead to poor health—e.g., through alcohol or tobacco consumption or mental illness.

It is not always clear that these negative outcomes are due to lower SEP rather than to other factors, such as geography. For example, poor children are more likely to suffer from pediatric asthma than nonpoor children, but only in urban areas (Aligne et al., 2000). On the other hand, Lauderdale, Thisted, and Goldberg (1998) found that regional differences in hip fracture rates could be explained by the state in which individuals initially qualified for their Social Security card and not where they currently lived. The authors hypothesized that this might reflect differences in nutrition earlier in life.

Education level is one component of SEP. Education embodies a concept of one's ability to process information, which includes the ability to read (literacy level) and understand complicated health and medical information. Higher education levels and better access to community and social networks can translate into better skills in getting quality care—engaging with physicians and other health providers and negotiating with health insurers. Education may also be associated with health-promoting behavior—for example, better understanding of human dietary requirements may translate into better eating habits. Those with less education or fewer skills are more likely to have jobs that require more physical labor and harsh working conditions, which could lead to worse health outcomes. Education also relates to income and wealth, as those with higher education levels tend to have higher incomes and wealth. Thus education is also indirectly related to the ability to afford health insurance and to afford higher quality health care.

Occupation, although closely related to other aspects of SEP, is itself an important component of SEP, separate from education, income, and wealth.

For example, clergy and teachers have relatively high education levels but relatively low wages, yet their occupations may represent leadership or social status above that of other occupations that command higher earning levels. It is also possible that there is a direct link between occupation and health. Some occupations that require physical labor may take a toll on physical health or may require greater physical stamina, which could act to either improve or harm physical health. Some occupations may lead to exposure to hazardous working conditions.

While the correlations between SEP and health are well documented (Acheson, 1998; Deaton, 2002; Kaplan and Lynch, 1997; Marmot, 2002; Smith, 1999; Sorlie, Backlund, and Keller, 1995; Turrell et al., 1999; Williams and Collins, 1995), the causal links between SEP and health are complex and can run in both directions. For example, poor health can affect employment, education, and occupation, which in turn can affect income and wealth (Covinsky et al., 1994; Wu, 2003); and conversely, those with lower income are likely to experience more stress and have worse health and less access to health care.

SEP and race and ethnicity are also interrelated. Most minority groups tend to have lower economic positions and less education than whites, on average. For example, the poverty rate in 2000 for non-Hispanic whites was 7.5 percent, compared with 22.1 percent for blacks, 21.2 percent for Hispanics, and 10.8 percent for Asians and Pacific Islanders (U.S. Department of Commerce, 2001a). Furthermore, median household income for non-Hispanic whites in 2001 was almost $46,000. This is lower than the median household income for Asians and Pacific Islanders ($55,521), but greater than the median household income for blacks ($30,439) and Hispanics ($33,447) (U.S. Department of Commerce, 2001b). There is considerable variation within each of the broad racial and ethnic populations. For example, within the Asian category, Laotians, Hmong, and Cambodians have higher rates of poverty and lower levels of household income than blacks or American Indians (U.S. Bureau of the Census, 1993). These three Asian subgroups also have higher rates of disability than other Asian groups and the white population (Cho and Hummer, 2000).[9]

Acculturation and Language Use

Acculturation is characterized as the dynamic bidirectional process whereby a person or group raised in another culture, typically immigrants,

---

[9]The Office of Management and Budget's recent revision of the racial and ethnic categories that calls for a new separate category for Native Hawaiians and other Pacific Islanders will provide for better tracking of the health of the Asian subgroup in the future.

come in contact with (in this case) the U.S. culture, resulting in subsequent changes in the behavior of both cultural groups. There are many components to acculturation. One component of the process is the degree to which an individual maintains ties to the culture of the country or region of origin—that is, the degree to which an individual holds norms, expectations, practices, and beliefs (in all areas of life, but for this report's purposes, with respect to health and health care) that are consistent with the culture of one's country or area of origin. Language use, particularly proficiency in understanding, speaking, reading, and writing English, is another component of acculturation. Place of birth, generation status, and time in the United States can also serve as indicators or components of acculturation.

U.S. culture itself is not homogeneous. Different people acculturate to different subcultures—for example, to a working-class or professional subculture, or perhaps to an ethnic subculture that has its own distinctive features. Furthermore, U.S. culture itself changes as the U.S. population changes—e.g., as new immigrant groups arrive and affect the areas in which they settle. For these reasons, acculturation can be difficult to define and measure because it is defined relative to a concept (U.S. culture) that cannot itself be precisely defined or measured. Rather, acculturation is marked by such factors as language use and proficiency, country of origin, or years or generations since immigration to the United States.

Acculturation affects health outcomes and interactions with the health care system. Various aspects of culture—behavior, diet, family environment—are related to health status and are transformed through acculturation. For example, adoption of a more typical U.S. diet could affect health positively or negatively, or perhaps both (e.g., less protein deficiency, but more obesity). Beliefs about health care may conflict with assumptions about the U.S. health care system; for example, views of illness as an imbalance and the use of traditional medical practices (e.g., herbalism) may be effective treatments for some conditions, but may also be inconsistent with more technically oriented U.S. medicine.

Many studies have shown associations of proxy measures of acculturation such as nativity (being foreign-born versus U.S.-born) or generational status with the health and health care of individuals (Clark and Hofsess, 1998; English, Kharrazi, and Guendelman, 1997; Sundquist and Winkleby, 2000; Crump, Lipsky, and Mueller, 1999; Guendelman and Abrams, 1995). Several studies suggest that acculturation is not always positively correlated with health outcomes. For example, less acculturation (more ethnic distinctness), as indicated by measures of nativity and measures of language proficiency, has been shown to confer a protective effect due to healthier lifestyles that are associated with greater family and social support and other forms of protective cultural practices. A greater degree of acculturation (as measured by indicators such as language proficiency—reading and writing in English; place of birth; and other variables) appears to be associated with a decline in some health indicators among Hispanics (Vega and Amaro, 1994). For example, more acculturated Latinos seem to experience early sexual initiation, higher rates of adolescent pregnancy, higher rates of low birthweight and infant mortality, and higher rates of hypertension and obesity. Highly acculturated Latinos are also more likely to smoke, use drugs, and consume more alcohol than less acculturated Latinos. Finally, breast-feeding is greater among less acculturated Hispanic mothers than among highly acculturated Hispanic mothers (Clark and Hofsess, 1998).

Language in particular can be a limiting factor in health care interactions and is often one of the biggest barriers to effective health care for immigrants. Discussing medical issues requires a rather advanced level of language proficiency that many nonnative speakers do not have, and the resulting lack of understanding can affect health and health care outcomes. For example, Vietnamese women who are fluent in English are more likely to have a routine place for health care and a regular provider than Vietnamese women who do not speak English (Yi, 1995). Spanish-speaking Hispanic patients were significantly less likely than non-Hispanic white patients to have had a physician visit, mental health visit, or influenza vaccination (Fiscella et al., 2002), whereas those who saw a Spanish-speaking doctor generally enjoyed better health (Perez-Stable, Napoles-Springer, and Miramontes, 1997). Frye (1995) described the example of an Asian family that brought their grandmother to an emergency room with a complaint of epigastric pain. Neither the patient nor any of the family members could explain the problem in English, so hospital personnel called an Asian American nurse and asked her to speak "Asian" to the family. But the Vietnamese nurse was unable to communicate with the Hmong family in their language.

Because disparities in health exist across different levels of acculturation, it is important to measure acculturation and language use to monitor, better understand, and target interventions to eliminate these disparities. Measures of acculturation and language may also be important covariates for understanding racial and ethnic disparities in health. Thus, gathering information about acculturation can help provide a better picture of a person's ethnic identity and its relationship to lifestyle behaviors that may affect the person's health.

THE IMPORTANCE OF UNDERSTANDING DISPARITIES IN HEALTH AND HEALTH CARE

Measuring and studying disparities serves several important functions. In this section, we highlight reasons why a better understanding of health disparities is needed to ensure the well-being of the overall population. In general, government health agencies at the federal, state, and local levels take on the responsibility of promoting and ensuring the health of the populations they serve. Because these government agencies are major funders of health insurance and health care services, they also have an interest in ensuring that these programs are administered fairly and are effective in meeting program goals. Other entities have an interest in promoting better health and health care as well. Better individual health may lead to more productive workers and less absenteeism, and thus employers have an interest in a healthy workforce. To the extent that healthy people and preventive health measures keep the costs of health care and health insurance low, employers, the entire health care industry, and the nation as a whole have an interest in promoting health among all their populations.

One reason it is important to study health disparities is to identify the problems and needs of specific groups of the population and in specific areas of the country. Measuring disparities can help in recognizing variations in the general health of populations and how those conditions change. Changes in health conditions over time and between groups may reveal areas that need further attention, such as the incidence and prevalence of particular health conditions and risk factors in population subgroups (see the paper by Fremont and Lurie in Appendix D). For example, Figure 2-2 shows trends between 1963 and 1994 in the percent of overweight children aged 6 to 11 by sex and race. The figure shows generally increasing rates of overweight status for all four groups—white males, black males, white females, and black females. However, rates have increased differently for each of these groups. In 1963 white males in the 6-11 year age range were overweight at higher rates than black males, but in the next 30 years the rate grew faster among blacks, who now have slightly higher rates of overweight status. White and black females initially were equally likely to be overweight, but the rate for black females has increased much more dramatically than for white females, such that in 1994, 17 percent of black females were overweight compared with 10 percent of white females.

Data on race, ethnicity, SEP, and language use and acculturation can be used to monitor how changes in the health care system and in the economy affect access to health care and to determine if these changes affect racial and ethnic groups differentially. For example, it would be useful to monitor and understand changes in health insurance coverage among different ethnic groups when health insurance premiums rise or unemployment increases. In short, early identification of problems can facilitate effective and timely interventions to eliminate disparities.

It is important to measure and study disparities in health and health care as part of the general goal of improving health and health care quality.

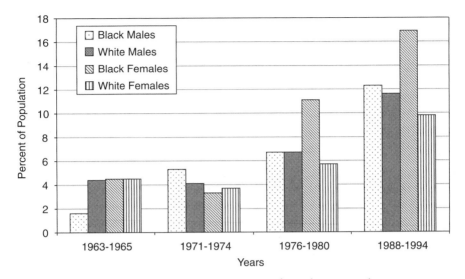

FIGURE 2-2 Overweight children 6 to 11 years of age, by race and sex.
SOURCE: Centers for Disease Control and Prevention, National Center for Health Statistics, National Health Examination Surveys and National Health and Nutrition Examination Surveys (NHANES).

Disparities may be an indicator of differences in either health care quality or health or both.

Since disparities in health and health care among racial, ethnic, language, and SEP subgroups have been identified, several agencies and governments responsible for ensuring the functioning of health care and public health systems have begun efforts to address identified disparities. The Healthy People 2010 initiative and the Health Disparities Report Card cited previously in this chapter are examples of two disparity-monitoring efforts by the federal government. These reports and others like them will provide periodic updates on the status of disparities and will help to hold accountable those agencies responsible for supporting the effective functioning of the health care system. In the case of racial and ethnic disparities, there is the important additional role of assessing and ensuring compliance with civil rights laws. While the existence of disparities in health and health care does not automatically imply discrimination, the monitoring of racial and ethnic disparities and of changes in disparities is critical to identifying potential problem areas and investigating the possibility of discrimination.

If the ultimate goal is to eliminate disparities in health and health care, then it is essential to understand the mechanisms that cause them. Measuring social variables such as race, ethnicity, SEP, language use, and acculturation and the extent to which these contribute separately and interactively to differences in health and health care is key to that understanding. In the next chapter, we discuss the measurement of these concepts of race, ethnicity, SEP, and language use and acculturation.

3

Measuring Race, Ethnicity, Socioeconomic Position, and Acculturation

The last chapter discussed the meaning of the concepts of race, ethnicity, socioeconomic position (SEP), and acculturation and language use and how they interact in affecting health and health care. This chapter discusses how these concepts are measured. The chapter first describes how race and ethnicity are typically measured in U.S. data systems and then describes measures of SEP and acculturation and language use. We then discuss some data collection issues that apply to the measurement of each of these concepts.

RACE AND ETHNICITY

The complexities of defining race and ethnicity make measuring these concepts complex as well. In part, these complexities arise because the concepts are defined socially and politically, and although categorizations are often made based on phenotypical characteristics, they are not clear biological concepts. Individuals classify themselves in racial and ethnic categories but are also classified based on others' perceptions. An individual's self-report of race or ethnicity is probably the most useful and the most consistent measure of his or her race and ethnicity and is, therefore, the one most frequently used. But even self-classifications may not be consistent across settings or time (Harris, 2002). Others' perceptions of an individual's race and ethnicity are less consistent but may be of interest in some circumstances; for example, a physician's assessment of a patient's racial or ethnic background may be relevant to understanding the pattern of treatment received.

The ways that race and ethnicity have been classified in data collection have changed over time and across settings. This includes classification in federal data collection systems, as we will detail below, but also data collected through record systems outside the purview of the federal government. This lack of consistency across settings or across methods for collecting data on race and ethnicity can pose problems for the interpretation of such data. For example, the population of American Indians is reported to have tripled between the 1960 and 1990 censuses, an increase that cannot be explained by migration or demographic changes (Sandefur et al., 2002) but that may be attributable to variations in self-identification or in the federal definition of that identity.

Office of Management and Budget Standards for Collecting Data on Race and Ethnicity

The federal government has been collecting data on race since the first U.S. census in 1790, with data on ethnic background added in later censuses. Census standards for racial classification have changed greatly over time. In the first census, enumerators classified free residents as white or "other," with slaves counted separately. Table 3-1 gives a history of how race and ethnicity have been classified in each census since 1790. Over time, as the nation's population became more diverse and as individual ethnic groups identified themselves, more categories were added and occasionally some were dropped. Some changes worth noting in recent decades include the addition of a question on Hispanic ethnicity as a separate item in 1977, and in 2000 the option for allowing individuals to identify with more than one racial group.

No federal standards for the collection of data on race and ethnicity existed until 1977, when the Office of Management and Budget (OMB) developed and issued a set of standards, called Statistical Directive Number 15, for the collection of these data. These standards were developed to provide consistency in defining race and ethnicity for civil rights legislative use, monitoring equal treatment, and other public policy uses (NRC, 2004). The Statistical Directive Number 15 classification system included four categories for race (white, black, Asian or Pacific Islander, and American Indian or Alaska Native) and two for ethnicity (Hispanic and non-Hispanic). Self-report was established as the preferred method of collecting data, and respondents were instructed to choose only one race and one ethnicity.[1]

---

[1]Because Hispanic origin is given special priority, equal to basic racial categories, in the OMB standards, the term *ethnicity* is often used to refer solely to the response to the question on Hispanic or non-Hispanic origin. Throughout this report, the term *ethnicity* is used both

TABLE 3-1 Racial Categories in the U.S. Census, 1790-2000

Year	Category
1790	Free whites, Other free persons, Slaves
1800 and 1810	Free whites; Other free persons, except Indians not taxed; Slaves
1820	Free whites; Slaves; Free colored persons; Other persons, except Indians not taxed
1830 and 1840	Free white persons, Slaves, Free colored persons
1850	White, Black, Mulatto
1860	White, Black, Mulatto, Indian
1870 and 1880	White, Black, Mulatto, Chinese, Indian
1890	White, Black, Mulatto, Quadroon, Octoroon, Chinese, Japanese, Indian
1900	White, Black, Chinese, Japanese, Indian
1910	White, Black, Mulatto, Chinese, Japanese, Indian, Other (plus write-in)
1920	White, Black, Mulatto, Indian, Chinese, Japanese, Filipino, Hindu, Korean, Other (plus write-in)
1930	White, Negro, Mexican, Indian, Chinese, Japanese, Filipino, Hindu, Korean, Other races (spell out in full)
1940	White, Negro, Indian, Chinese, Japanese, Filipino, Hindu, Korean, Other races (spell out in full)
1950	White, Negro, Indian, Japanese, Chinese, Filipino, Other race (spell out)
1960	White, Negro, American Indian, Japanese, Chinese, Filipino, Hawaiian, Part Hawaiian, Aleut, Eskimo
1970	White, Negro or Black, Indian (American), Japanese, Chinese, Filipino, Hawaiian, Korean, Other (print race)
1980	White, Negro, Japanese, Chinese, Filipino, Korean, Vietnamese, Indian (American), Asian Indian, Hawaiian, Guamanian, Samoan, Eskimo, Aleut, Other (specify)
1990	White, Black, Indian (American), Eskimo, Aleut, Chinese, Filipino, Hawaiian, Korean, Vietnamese, Japanese, Asian Indian, Samoan, Guamanian, Other Asian Pacific Islander, Other race
2000	White; Black, African American, or Negro; American Indian or Alaska Native (specify tribe); Asian Indian; Chinese; Filipino; Other Asian (print race); Japanese; Korean; Vietnamese; Hawaiian; Guamanian or Chamorro; Samoan; Other Pacific Islander (print race); Some other race (individuals who consider themselves multiracial can choose two or more races)

SOURCE: National Research Council (2004).

The use of these standards was required in all census and survey data collected by the federal government, as well as for federal administrative records and federally sponsored research (OMB, 1977). The standards did not apply to state or private data collection efforts except as required by

---

in that limited sense (to refer only to Hispanic origin) and in a broader sense to refer to other ethnic distinctions.

state-run federal programs, although some private surveys, especially those financed by the federal government, used these same classifications (see also Mays et al., 2003).

In the past 2 decades, the U.S. population has become significantly more heterogeneous as the populations of nonwhite minority groups have grown. After congressional hearings in 1993 (House Subcommittee on Census, Statistics, and Postal Personnel), OMB announced a review of the 1977 standards. In addition to accounting for increased heterogeneity, the review was also to address challenges in identifying those of multiracial heritage. The interagency review included tests of questions regarding race, multiple racial categories, and ethnic categories in a special supplement to the 1995 Current Population Survey (CPS) (BLS, 1995), as well as in both the 1996 National Content Survey and the Race and Ethnicity Targeted Test (U.S. Bureau of the Census, 1996).

Based on the results of these tests and public comment, OMB revised Statistical Directive Number 15 in 1997. The revised standards featured three important changes. First, five racial categories were to be used in measuring race: Black or African American, White, Asian, American Indian and Alaska Native, Native Hawaiian and other Pacific Islander. Second, respondents would be permitted to select more than one race. And third, the question on ethnicity was to be changed by asking respondents whether or not they were Hispanic or Latino, and was to be asked before the race question was asked.

The standards were to be effective immediately for all new and revised federal systems and no later than January 1, 2003, for existing systems. Agencies are permitted to add categories when more detailed data are needed as long as the data can be aggregated to the minimum five categories of race. As with the 1977 guidelines, these minimum standards apply to all federal data collection activities but not to state or private-sector data collection, except when required for federally sponsored statistical data collections, including all federal administrative and grant reporting (OMB, 1997).[2] OMB emphasized "The categories represent a social-political construct designed for collecting data on the race and ethnicity of broad population groups in this country, and are not anthropologically or scientifically based" (OMB, 1997, p. 16).[3]

In many instances, especially in health settings where culture and environment play an important role in outcomes, finer measures of ethnicity are needed in order to measure heterogeneity within broad ethnic categories. For example, heterogeneity among Hispanics in terms of their health has

---

[2]See http://www.whitehouse.gov/omb/fedreg/ombdir15.html.

[3]As social and political changes occur, these broad categories may or may not be appropriate in the future. Presumably the OMB will revisit the categories as the need arises.

been documented. One landmark study of low birth weight and infant mortality among Hispanics (Becerra et al., 1991) showed that infant mortality rates varied by the infant's mother's Hispanic ethnicity: infants of Puerto Rican descent had the lowest birth weights and the highest infant mortality rates of all Hispanic groups, while infants of Cuban descent had the lowest infant mortality rate. Thus, especially for more localized use of data, it is important to distinguish among ethnic groups within a racial category (e.g., between Filipinos and Japanese). State and local public health agencies need to target programs and interventions to particular groups with specific problems and communication needs. More refined measures of ethnicity are also needed because these ethnic groups want to know about the health of the populations in their communities. The revised OMB standards allow for the collection of data in more narrowly defined categories such as these, as long as the additional categories can be aggregated back to the standard categories.

The Importance of Collecting Data on Race and Ethnicity

As discussed above, data on race and ethnicity are necessary to measure disparities in health and health care in order to understand the causes of them. Such data can be obtained from federal, state, and private-sector data collection systems, including surveys and records used for health care services and programs. There is currently considerable variability in the kinds of data collected in these systems, and the panel believes that it is important for each system to collect standardized racial and ethnic data. A concerted effort involving both the public and private sectors is needed to improve the recording of individuals' racial and ethnic descriptors in health information systems and to enhance the nation's capacity to generate health information of comparable content and quality for all racial and ethnic groups, particularly those segments at highest risk of health problems.

CONCLUSION 3-1: Measures of race and ethnicity should be obtained in all health and health care data systems.

SOCIOECONOMIC POSITION

Chapter 2 identified crucial dimensions of SEP—education, occupation, current income, wealth, and life history of income—and briefly discussed how they are related to health and health care. In this section, we briefly discuss measurement of these dimensions of SEP. There is a significant literature on the measurement of these dimensions and with respect to their relationship to health and health care (see the paper by O'Campo and

Burke in Appendix C; Duncan et al., 2002; Oaks and Rossi, 2003; Williams, 1996).

Educational Attainment

Level of education reflects several aspects of resources and ability relevant to health and health care. In addition to conveying a certain level of ability—intellectual, behavioral, and financial—it can also be used as a proxy for one's knowledge of or ability to process and understand factors affecting health as well as diagnoses and treatments. It may also bear on one's ability to communicate effectively with physicians and other health care professionals.

One important characteristic of educational measures is that they are relatively stable over the adult life course and thus measure a "permanent" component of SEP. Thus education is also a measure of resources that have consequences throughout the life course. Education is correlated with other concepts of SEP—occupation, income, and social status—that are related to health and health outcomes.

Most often, education level is measured by years of schooling completed or by credential obtained (for example, high school diploma, associate's degree, bachelor's degree, etc.). Usually, for health surveys and data collection, level of education is measured in one question. The federal government currently has no standard way of measuring educational attainment in its data collection efforts, but a Federal Interagency Committee on Measures of Educational Attainment recommended that the measure of education used on the 2000 census long form should be the model for measuring education in all federal surveys and administrative data collections. This measure, shown in Box 3-1, combines years of schooling completed through high school with detailed categories for college and advanced degrees (Federal Interagency Committee on Measures of Educational Attainment, 2000).

One deficiency of measures of educational attainment is that they do not reflect geographic and individual variations in the quality of education and therefore imprecisely reflect literacy levels and other intellectual skills.

Occupation

Occupation constitutes another distinct aspect of SEP. A limited amount of information on occupation can be collected by a single open-ended response item. This measure has the advantage of simplicity and brevity, and so can be collected on record systems where there is not much time to get many details about one's occupation. Furthermore, people are less reluctant to report their occupation than their income and wealth. However,

---

BOX 3-1
Educational Attainment Question
from the 2000 Census Long Form

What is the highest degree or level of school this person has COMPLETED?
Mark ONE box. If currently enrolled, mark the previous grade or highest degree received.
☐ No schooling completed
☐ Nursery school to 4th grade
☐ 5th grade or 6th grade
☐ 7th grade or 8th grade
☐ 9th grade
☐ 10th grade
☐ 11th grade
☐ 12th grade, **NO DIPLOMA**
☐ **HIGH SCHOOL GRADUATE**—high school DIPLOMA or the equivalent *(for example: GED)*
☐ Some college credit, but less than 1 year
☐ 1 or more years of college, no degree
☐ Associate's degree *(for example: AA, AS)*
☐ Bachelor's degree *(for example: BA, AB, BS)*
☐ Master's degree *(for example: MA, MS, MEng, MEd, MSW, MBA)*
☐ Professional degree *(for example: MD, DDS, DVM, LLD, JD)*
☐ Doctorate degree *(for example: PhD, EdD)*

SOURCE: U.S. Bureau of the Census (2000).

---

coding these responses can be quite difficult. In addition, such a measure may not provide much information on social or economic conditions gained through the occupation, or on environmental and work conditions to which one might be exposed to while at work. Thus, the simple measure may not always be a strong indicator for health research. Additional information on specific occupation, industry, and years at the position may be collected to give a better sense of the conditions of the occupation. Studies concerned with environmental and occupational risks that may affect health would need additional information on those conditions. These are more difficult to collect and unlikely to be collected except for studies examining very specific health threats.

The federal government's standard categorization system for measuring occupation, the Standard Occupation Classification System (see www.bls.gov/soc/home.html), hierarchically classifies occupations into both broad and very specific groups. The system is intended to be used by all federal statistical agencies that collect data to classify workers.

Current Income

Current income can be an important determinant of the resources available to an individual or family. It is usually measured as an individual's, family's, or household's total cash income, over a month, the preceding 12 months, or the calendar year, and usually on a pretax basis. Posttax information may be a more relevant measure, but it is more difficult to obtain (Duncan et al., 2002).

Individual, family, and household income measures can be relevant to health and health care. Family or household income is usually based on the sum of the incomes of family or household members and is often used to measure individual SEP, implicitly assuming that resources are shared by the family. In reality, however, some members may have access to more of the family's resources than others, or have more control over how the assets are allocated; for example, when one adult family member is a homemaker and one has paid employment. Conversely, individual income may not accurately measure an individual's resources if those resources are shared with family members. When family income is measured it is also important to know the number of family members and sometimes the composition of the family (in terms of adults and children or perhaps elderly adults as opposed to nonelderly adults) to determine the adequacy of the resources.

Measures of income can be considerably affected by how questions are asked, as shown by comparisons of results from single-item income measures like the one in the long form of the census and the more complex series of items in the Annual Demographic Supplement to the CPS. Furthermore, many respondents consider income a sensitive topic, leading to high levels of item nonresponse on income items in surveys and some administrative systems.

Wealth

Wealth measures a dimension of SEP different from current income. The two are correlated, but the correlation is not perfect.[4] Income is a measure of current flow of resources, although higher levels of income may enable a family to accumulate wealth. Wealth reflects an individual's or household's stock of accumulated resources, such as property, savings, and other assets at a point in time. Since wealth is transferable from generation

---

[4]For example, elderly people often have little current income but significant wealth, and so wealth may be more important than income when assessing their SEP. In contrast, younger people may have high current income, but lower levels of accumulated wealth. Venti and Wise (1999) find substantial variation in wealth among people with similar income from earnings, even among those of the highest wealth levels.

to generation, family wealth can be an important factor, although it may be very difficult to measure. Wealth is important because it may be a buffer against periods of low income or high consumption (such as a catastrophic health event requiring expensive health care that is not covered by insurance), and thus may affect both health and health care. For example, by buffering against such periods, wealth may allow an individual or family to avoid the effects of deprivation, which may result in greater health. Duncan and colleagues (2002) and J. Smith (1999) both document the significant effect wealth has on health even when other measures of SEP are considered. Wealth may also allow a family to obtain health care treatments that could not otherwise be paid for through current income resources. For example, the decision to obtain care may depend on whether accumulated assets are available (and sufficiently liquid) to pay for the procedure.

The inequality in wealth levels between blacks and whites is even more pronounced than the inequality in income (Barsky et al., 2002; Oliver and Shapiro, 1995). Differences exist both within levels of social class—middle-class blacks have lower levels of net worth and net financial assets than middle-class whites—and among those with similar educational background—college-educated blacks have lower levels of net worth than college-educated whites (Oliver and Shapiro, 1995).

Wealth is most effectively measured by collecting extensive data on the value of financial assets (e.g., savings accounts, stocks, and bonds), retirement accounts such as 401(k) funds, pensions, real estate holdings and home ownership status, business equity, and ownership of large durables, such as vehicles. Debt information is also part of wealth measurement and may be used to capture a household or individual's net wealth. Extensive measures of wealth are collected in only a few surveys, such as in the Health and Retirement Survey. In other surveys, less extensive measures of wealth are collected. For example, the Medical Expenditure Panel Survey (MEPS) collects information on asset and debt levels, and the National Health Interview Survey (NHIS) and National Health and Nutrition Examination Survey (NHANES) collect data on home ownership. Measures of wealth are rarely found in administrative or private records unless they are needed to administer the program (e.g., Medicaid collects some information on assets). O'Campo and Burke (see their paper in Appendix C) review how wealth is measured in health and health care data collection systems.

Lifetime Income and Wealth History

While wealth and current income measure current economic resources, lifetime income and wealth and their dynamics can also be important dimensions of SEP. For example, prolonged economic deprivation, which may entail exposure to negative environmental factors, poorer nutrition, or stressful conditions, may wear on one's health. Exposure to economic deprivation during critical times over the life course may affect health. Or severe fluctuations in current income or wealth levels may have negative impacts on general health.

Measures of lifetime income and wealth are usually available only from longitudinal surveys and quite difficult to collect. The Panel Study of Income Dynamics is unique in that this longitudinal survey has followed the same families since 1968 and has collected extensive income and wealth data on these families. Another possible source of lifetime income measurement is from earnings records reported to the Social Security Administration (SSA), which could be linked to Medicare records to measure lifetime earnings for beneficiaries.

Area-Based Measures of SEP

A person's address and Zip Code are obtained for many data systems and are often used to link area-based measures of SEP to the individual record as a proxy for the SEP of the individual. For example, per capita income for the Zip Code code area in which the individual resides may be used as a proxy for an individual's income, or the percent of the residents of the Zip Code area with a college degree may serve as a proxy for individual education level. These area-based measures of SEP usually come from U.S. Census long-form data items, which include information on income, employment status, education, language, country of origin, citizenship, and housing and residency. The smallest area of geography for which long-form data are released is the census block group, but data are also released at the tract and Zip Code levels.[5] For example, Singh and colleagues (2003) linked county-level and census tract-level information on poverty rates and matched them to cancer registry data (from the Surveillance, Epidemiology, and End Results program) on incidence and mortality from various forms of cancer to study the relationship between cancer and socioeconomic position. More commonly, however, Zip Code-level information (e.g., median income in the Zip Code) is used as a proxy.

The process of matching an individual's address to a census unit (tract or block group) or other geographic unit (a county) is called geocoding. Geocoding allows the linking of individual data to group level data from different sources. We discuss the use of geocoding for developing proxies for individual measures later in this chapter.

---

[5]A census block group is a set of census blocks that optimally includes 1,500 people, but sizes of block groups in the 2000 census ranged from 300-3,000 people. Tracts optimally include 4,000 people but in the 2000 census ranged from 1,000-8,000. Zip Code areas average 30,000 people.

Collecting SEP Data

Many health surveys and program data collection efforts do not collect detailed SEP data (Duncan et al., 2002; O'Campo and Burke, Appendix C). Measuring SEP in general is as difficult as measuring race and ethnicity. Many questions and follow-up questions may be needed in order to accurately measure income and wealth. Income and wealth are also highly sensitive topics for questionnaires. People tend to underreport their income and are often reluctant, for confidentiality reasons, to report it at all. The depth of this problem depends on the survey and on the type of income information requested. For additional reading on the quality of income and wealth data, see Bound and Krueger (1991); Cilke (1998); Coder (1992); Hotz and Scholz (2002); Moore, Stinson, and Welniak (1997); Rodgers, Brown, and Duncan (1993); Roemer (1999 and 2000).

Because of the difficulty of measuring income and wealth, many health surveys do not obtain detailed information on those dimensions of SEP. Some surveys (e.g., NHANES, NHIS), sensitive to respondent reluctance to provide detailed information on income and wealth, only ask respondents to record the amount of income they receive each year on a pretax basis and also to check a category of annual income level. On the other hand, education and occupation are less sensitive items on surveys and useful data can be collected with single questions.

Some administrative databases collect detailed information on income if a certain level of income is a prerequisite to participate in the program. For example, income information is collected to determine eligibility for the Medicaid program and some states with assets tests collect information on household assets such as savings accounts, investments, and automobiles.[6] Some surveys collect information about participation in low-income programs such as Medicaid, State Children's Health Insurance Program (SCHIP), the Women, Infants, and Children (WIC) program, and food stamps. Thus, participation in these programs can be used as a proxy measure of low-income status. However, the income eligibility limits of these programs vary across programs and across states. Furthermore, individuals tend to underreport their participation in these programs. Otherwise, the collection of SEP data in administrative records systems is rare and often limited to measures of education and occupation.

---

[6]Measures of assets from these records may understate their true value because applicants to means-tested programs like Medicaid have an incentive not to report all of their income and assets. Survey measures of means-tested program participation and benefit receipt are also underreported by survey respondents such that measures of income from surveys may be biased downward in surveys as well (see Hotz and Scholz, 2002; Wheaton and Giannarelli, 2000).

SEP indicators have a substantial relationship to health and health care and can be important mediating factors for understanding racial and ethnic disparities. The panel therefore recommends that collection of these measures in health surveys and administrative data collections be made a priority, both to better understand racial and ethnic disparities and as a means of identifying effects on deprived groups that are not defined by race or ethnicity. Because of limitations in survey lengths and in the data that can be collected as part of an administrative process, it will not be feasible to collect a full set of SEP measures in all instances.[7] However, DHHS should consider which measures can be collected in different settings and push for their collection.

CONCLUSION 3-2: Measures of socioeconomic position should, where feasible, be obtained along with data on race and ethnicity.

ACCULTURATION AND LANGUAGE USE

Language barriers and cultural differences between patients and providers and other actors in the health care system can lead to lower quality of care or poor health outcomes.[8] With knowledge of the acculturation of a patient or a community, health care providers are better positioned to provide effective health care, and the causes of disparities in health and health care can be better understood.

As discussed in Chapter 2, acculturation involves a complex process of interaction with aspects of the U.S. culture by individuals from different cultures. Less acculturated individuals will be more likely to have been born outside the United States, to speak their language of origin, and have cultural traits that are more closely linked to their culture of origin. An individual's degree of acculturation can be measured by cultural characteristics, such as language preference and English proficiency or cultural practices, or by other variables that measure status, such as place of birth or years or generations living in the United States. These variables may yield more information useful for studying health and health care because they are closer proxies to factors that tend to vary across cultures that may affect

---

[7]In the context of this broad review of data systems and their collection of SEP, the panel cannot specify which measures of SEP should be collected in every data system. In later chapters, where specific data systems are discussed, the panel suggests measures of SEP it believes may be reasonable to collect. The primary point is that greater effort is needed to collect SEP data systematically along with data on race and ethnicity.

[8]In general, language barriers occur when the patient and the health care professional cannot understand each other. In most cases in the United States, language barriers occur when one party (most often the patient) is not fluent in English.

health (for example, diet or views of traditional medicine) and health care (interpretation and understanding of health care treatment). In an ideal setting a battery of questions would collect extensive measures of acculturation. However, this is often not possible. Status variables are more easily collected, and so are often used as proxies for the degree of acculturation—presuming that individuals born in another country tend to be less acculturated than those born in the United States, or that those who have lived in the United States for a longer time are more acculturated than those who have arrived more recently. This conceptualization of acculturation, although simplistic, provides the opportunity to characterize this phenomenon with the use of individual cultural and social proxies (Berry, 2003).

Language proficiency (in most instances in the United States, this means the ability to communicate in English) is a fairly reliable proxy measure of acculturation and can be assessed in many different ways—ranging from simple to fairly complex. Longer language acculturation scales have been developed as instruments that assess acculturation processes with more multidimensional domains, including language and other indicators such as generation status and cultural orientation variables. The Acculturation Rating Scale for Mexican Americans-II (ARSMA II) is a good example of this type of multidimensional scale that includes concepts such as integration, separation, assimilation, and marginalization (Cuellar, Arnold, and Maldonado, 1995).[9] Language indicators within these multidimensional scales have been shown to be powerful predictors of health status among Hispanics (Deyo et al., 1985; Cobas et al., 1996). Extensive scales of language proficiency or of acculturation exist and can be used when more detail is needed (Cuellar, Arnold, and Maldonado, 1995), but because of their length, are better suited for surveys than for records-based data collections.

Some measures of language proficiency may include a set of questions on language that are scaled to give a score of language proficiency. One short language-based acculturation scale (Deyo et al., 1985) has just four questions: What language do you prefer to speak? What language is most often spoken in your home? What was your first language as a child? Do you read any English? These four simple questions cover a range of possibilities of language use and offer a useful gauge of a person's English fluency. But often only one question is asked about the respondent's preferred or primary language. These simple scales of acculturation based on language proficiency, either combined or as a single question, have proven to be good and reliable indicators for health care research among immi-

---

[9]Acculturation scales for Asian subpopulations have also been developed and are being used to assess variations in health and health care among these groups (Anderson et al., 1993).

grant and racial and ethnic minority populations (Deyo et al., 1985; Cobas et al., 1996; Aguirre-Molina, Molina, and Zambrana, 2001; Byrd, Balcazar, and Hummer, 2001; Unger et al., 2000).

Language proficiency does not convey the entire spectrum of acculturation as a complex construct, but it has proven to be a reliable measure. Language generally accounts for over 70 percent of the variance in the total acculturation score on acculturation scales (Clark and Hofsess, 1998).

In addition to language, other elements of acculturation experience include: place of birth, cultural expressions and feelings, attitudes, emotional behaviors and beliefs, and ethnic loyalty (Cuellar et al., 1995; Marin et al., 1987; Balcazar, Peterson, and Cobas, 1996; Clark and Hofsess, 1998).[10] Place of birth is also a simple and useful (although not perfect) proxy indicator of acculturation. Such information is likely to be useful to improve program administration, to target language—appropriate educational materials and information, to suggest providers, or to allocate translation services. Generational status and time in the United States are additional indicators of the acculturation experience and can provide pertinent information about health care access and utilization of health services in addition to differences in health outcomes among immigrant populations living in the United States. However, these indicators, along with language, do not satisfy the criteria of multidimensional acculturation measurement (Clark and Hofsess, 1998; Chun et al., 2003).

These are not perfect measures, of course. An individual's degree of acculturation is dependent on other characteristics as well; home, work, and social interactions, exposure to the cultures in the United States, or proficiency with the English language before immigration. Other aspects of the individual that are associated with acculturation include socioeconomic status, discrimination, occupational experiences, and neighborhood environments (Clark and Hofsess, 1998; Cuellar et al., 1995). Furthermore, some of these measures, such as place of birth, will not be relevant to some Native American populations for which language and cultural practices may be barriers to interactions with the health care system. An even more challenging concept for acculturation is the notion of who represents the host majority culture. Many "minority" communities represent the majority or predominant population in a given geographic area or community. However, discussion of the implication of this for acculturation is beyond the scope of this review.

---

[10]In a health care treatment setting, in addition to collecting data on the language preference of the patient, it may also be useful to collect data on whether any health care professionals treating the patient spoke the language of the patient.

CONCLUSION 3-3: Measures of acculturation and proxies such as language use, place of birth, and generation and time in the United States should, where feasible, be obtained.

We noted earlier that race and ethnicity are fluid concepts. While the meanings of race and ethnicity change over time, the relationship between race and ethnicity and SEP and acculturation may also change over time. For example, if patterns of immigration change (i.e., new groups from new areas immigrate to the United States and groups that have immigrated continue to assimilate), the existence and nature of disparities may also change. It is also possible that the economic and social positions of minority groups may change. Thus, how race and ethnicity are measured and the relationship of these two concepts to social and economic position and acculturation are likely to be reconsidered from time to time.

CROSS-CUTTING ISSUES IN MEASUREMENT AND USE OF THESE DATA

Some measurement issues are common to the collection of data on race, ethnicity, SEP, and acculturation and language use. Survey and administrative data systems are often limited in the amount of data that can be collected on each of these dimensions. As indicated earlier, sample sizes also limit statistically reliable estimation of disparities in health and health care within small population subgroups. In addition, health data are collected for several different purposes. Some data are collected to administer a program; for example, the primary use of Medicare claims data is to process payments for services. Data collected in the application process for health insurance are used to underwrite policies. Data on race and ethnicity are sometimes used to enforce civil rights laws. Because many of these data sets are not collected for research purposes, they may not have all the characteristics of an ideal data set for research on disparities in health and health care. Confidentiality and privacy issues may also limit their use because individuals who provided the data may not have been informed or have consented to let their data be used for purposes other than the primary reason the information was collected. Finally, because no one database is fully comprehensive for measuring race, ethnicity, SEP, and language use and acculturation, data linkages are often necessary to avoid the cost of new collections. The linkages can be difficult to arrange and come with their own privacy and confidentiality problems. This section discusses some of these problems in the collection of data on race and ethnicity, SEP, and language use and acculturation.

Data Content and Coverage

Surveys

A number of surveys collect information on health, health care access, utilization, and quality. Some surveys are conducted at the national level, often sponsored by the federal government. Some surveys ask specifically about health status, such as the National Health Interview Survey, while others focus on other topics such as medical expenditures (the Medical Expenditure Panel Study) and collect only limited information on health status.

States and localities conduct their own surveys as well. For example, the state of Hawaii has conducted the Hawaii Health Survey since 1968. The Pregnancy Risk Assessment Monitoring System (PRAMS)—state-based but coordinated by the Centers for Disease Control and Prevention (CDC)— surveys a sample of women in 13 states and the District of Columbia who recently gave birth to collect information on maternal behaviors before, during, and after pregnancy.

Most of these surveys collect data on race and ethnicity.[11] In federally supported surveys, if racial and ethnic data are collected, the minimum OMB standards for the collection of these data must be followed. There are, however, no such standards for the collection of SEP data (although there is a standard way to classify occupation), nor for data on language use and acculturation. Because these surveys focus on health data collection, they do not contain extensive measures of SEP. (See the paper by O'Campo and Burke in Appendix C for a complete listing of what SEP measures are collected in these surveys.) Questions about educational attainment and occupation are often included in the surveys, but only very limited information on income and wealth is collected.

Most national surveys are designed to produce estimates for the nation as a whole, not specific racial and ethnic subgroups. Sample size limitations in many studies allow reliable estimates of health status or health care utilization for only the larger racial and ethnic groups—whites, blacks, and Hispanics. Some surveys do oversample[12] certain minority groups (e.g., the

---

[11]See the DHHS Directory of Health and Human Services Data Resources: http://aspe.hhs.gov/datacncl/datadir/index.shtml.

[12]A specific population is oversampled when a survey interviews a disproportionately larger number of units (e.g., households or individuals) of that population than they constitute in the total population being sampled; for example, a survey that oversamples Hispanics will attempt to interview a higher proportion of Hispanics in the survey than Hispanics represent in the overall population. This is usually done in order to create a sample size of the specific population that is large enough to make statistically reliable estimates of the characteristics of the population.

Health and Retirement Survey (HRS) and NHIS both oversample Hispanics and blacks, and the NHANES oversamples blacks and Mexican Americans). But samples are often not large enough to produce reliable estimates for smaller racial groups—Asians, Native Hawaiians and Pacific Islanders, Alaska Natives and American Indians. An exception is the Hispanic Health and Nutrition Examination Survey (Hispanic HANES), conducted from 1982 to 1984 for three Hispanic subgroups, Cubans, Puerto Ricans, and Mexicans. The examination of differences in health status among Hispanic subgroups and among non-Hispanic whites and blacks and Hispanics was made possible with this survey.

Special efforts, like the Hispanic HANES, are often required to obtain adequate coverage and sample size for specific subgroups. A survey, unlike administrative records, affords the researcher more control over the populations that are covered (although limitations in survey methods may limit coverage). But with limited resources for data collection, it is not always possible to conduct special surveys with the same frequency as those conducted for the nationally representative populations. Thus, which populations are covered in surveys is determined mainly by resources and other government priorities (e.g., the need to obtain answers to specific policy questions) rather than by limitations inherent to the data collection methods and processes.

Administrative Data Systems

Data collected through federal and state health programs or through the operations of health care providers (e.g., hospitals) are often used to measure and understand health status and health care utilization. Examples of administrative data systems include the Medicare Enrollment Database, which represents enrollees since 1966, and the federal Healthcare Cost Utilization Project, which collects hospital discharge abstracts in a uniform way from 28 state data organizations. These administrative data sets are more often used for their information on health care utilization rather than on health status, as measures of the latter are not usually collected as part of the administrative process. Administrative data are also used to measure disease incidence; for example, the CDC maintains a number of disease surveillance reporting systems to which hospitals, labs, clinics, and other health care institutions submit information on disease incidence (for example, bacterial meningitis, HIV/AIDS, food-borne illnesses).

Administrative data are collected to fulfill the particular purposes of a program or administrative process. Therefore, unlike survey data, where content is often (within budget and time constraints) under the control of the data collection agency, the content of administrative data is usually limited to the specific information needed for administrative objectives. For example, a clinic may ask for a patient's age, residence, occupation, and insurance coverage as well as medical information as part of the admission process. Similarly, information on income and assets might be collected to assess eligibility for Medicaid. It is sometimes possible to collect additional data items, but usually only a limited number. As we will discuss in Chapters 5 and 6, race and ethnicity are sometimes included, although not consistently.

The populations included in administrative data collections are limited both by the scope of the specific program and to those directly served by the program. For example, only Medicare enrollees are covered by the Medicare Enrollment Database, and only those with a specific disease are covered by the disease surveillance systems. An advantage of administrative data sets, however, is their large sample sizes, which make it possible to obtain measures of interest for relatively small groups. Furthermore, medical records and billing or reimbursement systems contain extensive information on health care received by individuals. Thus, with the addition of suitable items on race, ethnicity, SEP, and acculturation, these data could be highly useful for studying disparities in health care.

Statistical Uses of Data

Information about individuals is used to make general statistical inferences about populations to monitor trends in disparities, to understand how disparities arise, and ultimately to design interventions to eliminate and reduce them. These inferences may be descriptive—describing health status, disease prevalence, and health care outcomes for a population of interest—or they may be used to draw causal conclusions about why an outcome or a difference in outcomes between groups is observed. While information at the individual level is needed in order to make these inferences, the specific identities of individuals are irrelevant and inferences are always drawn at an aggregate level.

Such statistical uses are distinct from other uses of the data that require information about a specific individual. For example, income data on individuals applying for Medicaid are collected to assess eligibility for the program; data on particular hospitals or individuals treated in hospitals are collected to ensure enforcement of civil rights laws. Data on individuals may be used to underwrite insurance policies—that is, to assess whether or not coverage should be offered to an individual and at what rate. In each of these cases, data on individuals are collected to take action regarding a specific individual. In contrast, data on individuals used for statistical purposes are collected to make inferences at an aggregate level.

Using data on individuals can create a situation where an individual's identity and private information could be disclosed and could potentially be used in a way that harms the individual. To prevent this from happening, when an investigator gains access to a confidential data set, he or she is usually required to agree to use it for statistical purposes and not release any confidential data on an individual. Breaches of such agreements may be punished by loss of access to data, of research funding, and of permission to conduct research.

Many of the data sources used to understand health disparities are not collected specifically for these statistical purposes, but rather are used to administer services and programs. Their use for statistical purposes is secondary. The panel, in this report, will make recommendations that encourage the collection of additional items of race and ethnicity, SEP, and language and acculturation where possible so that statistical inferences about disparities can be made. But it does so with the recognition that these data need to be useful to the federal, state, and private institutions and systems for which they are collected, and ultimately for the individuals who provide the data voluntarily, not just for statistical purposes. There are some examples of how data on race and ethnicity could be used to benefit the individuals who provide the data and the institutions who collect the data. For example, a health insurance plan might collect ethnicity and language data on enrollees to target culturally appropriate information or program interventions to individuals in their primary language. Or a plan may want to target information on disease prevention to enrollees who belong to racial or ethnic groups with higher prevalences of certain diseases.

CONCLUSION 3-4: Health and health care data collection systems should return useful information to the institutions and local and state government units that provide the data.

Data Linkages

Data linkages, or combining variables from two or more data sets, can facilitate new analyses (for policymaking, quality improvement, and research) without the expense and time needed for additional data collection. While there are tremendous opportunities for new analyses with linked data, there are also barriers to linking data sets.

Linking data sets usually entails bringing together information that identifies individuals, such as names, social security numbers, or a program identification number. This means that privacy and confidentiality regulations and concerns must be addressed. Confidentiality concerns are also increased when data are linked because more than one data system is being employed, and individuals who provide data to one system may not want the information made available to another.

Sometimes there are legal limitations on the use of data linkages. For example, employers are legally forbidden to link some claims data from their employees' health insurance records to employer-based records without protection of the employee's privacy. The Health Insurance Portability and Accountability Act (HIPAA) regulations require removal of identifiers from publicly released data, although exceptions can be authorized (with appropriate safeguards) when required for research. The use of social security earnings data by researchers outside the agency is severely limited.

A further barrier to data linkages is the need for negotiation across agencies and entities that maintain the data, which might have varying confidentiality provisions. Thus, linkages across agencies may require complex interagency negotiations.

Several methods are used to guard against harmful uses of linked data and to protect confidentiality. Masking and deidentification are two procedures that maintain the integrity of an individual's data but strip any personally identifying information from the linked record. The National Center for Health Statistics (NCHS), the Agency for Healthcare Research and Quality (AHRQ), and the Census Bureau all maintain restricted-access data centers that are housed in a secure setting but make the data available to researchers with proper credentialing and assurances of nondisclosure. These techniques facilitate the use of linked data. (See NRC, 2000 and NRC, 1993 for more extensive discussions of these methods.)

Sometimes when it is impossible to link data on individuals from two or more data sets, individual data from one set are linked to geocoded area-based measures from another set of data, which serve as a proxy for individual measures. As mentioned previously in this chapter, geocoding and the use of area-based measures are not perfect proxies for an individual-level variable. Area-based measures both at the Zip Code and census tract level are not as precise as individual-level data (Geronimus and Bound, 1998). But some area-based measures have been found to be better than others for health outcomes models: for example, aggregate income, education, and occupation were better predictors of health outcomes than socioeconomic index measures (Geronimus and Bound, 1998). This study also found that census tract-level measures are not significantly better than Zip Code-level measures. Krieger (1992) found that block group measures of SEP performed better than census tract measures of SEP for some health outcomes, but that the opposite was true for others. In a more recent study that examined many different health outcomes (e.g., birth and death outcomes, incidence of cancer and other diseases, and homicide), Krieger et al. (2003) found that census tract- and census block-level measures of SEP gave consistent parameter estimates of the effects of these SEP measures on outcomes across different racial, ethnic, and gender groups, while Zip Code-level measures were less consistent. This study also found that the percent of the area in poverty (tract, block, or Zip Code) was the SEP area-based measure that gave the most consistent estimates over different racial, ethnic, and gender groups than other area-based measures of SEP.

Linking across data sets has great potential payoff in terms of increased content coverage over a single source of data. For example, SEP and language data from other data sources could be linked to a data set that does not cover these items, or demographic information or information on health outcomes could be linked to information about health care received. While there may be technical and privacy issues to contend with in the linkage of these data, these issues can be addressed. Linking data sets collected by entities outside DHHS (e.g., by states and the private sector) may be more difficult because data sharing agreements may need to be negotiated and because data formats may be less consistent. However, since many of the data sets available to measure health disparities are within DHHS, some of the burdens of dealing with department cross-agency protections of privacy and confidentiality could be reduced if strong leadership is exercised by the department's groups and agencies with data collection and coordination responsibilities.

CONCLUSION 3-5: Linkages of data should be used whenever possible, with due regard to proper use and the protection of confidentiality in order to make the best use of existing data without the burden of new data collection.

Improvement of the data systems available to study racial and ethnic disparities in health will impose some additional burdens and costs on data systems. As we stated in the introduction to this report, it is beyond the panel's charge and would require a special set of expertise to provide a detailed assessment of the costs of these improvements. However, some general principles for collecting costs and reducing burden are discussed in the next chapters of this report. The collection of data on race and ethnicity and some simpler measures of SEP and acculturation and language has proven to be feasible and not difficult, although some of the more complex measures of SEP and acculturation and language use are difficult to collect and may not be practical to collect in every situation. It is true that the collection of these data may require some changes in computer systems, but such changes occur in the normal course of events from time to time and would not be more burdensome in this case. The major costs in collecting these data appear to be at the point of contact—that is, when a patient's or program enrollee's information is obtained. These burdens can be reduced, for example, by designing the systems in a way that avoids repeatedly collecting the same information.

4

DHHS Collection of Data on Race, Ethnicity, Socioeconomic Position, and Acculturation and Language Use

The Department of Health and Human Services (DHHS) is a major producer of data used in health and health care research. Through its survey data collection activities and the administration of its programs, the department collects an enormous amount of data that is used to study disparities. In this chapter, we provide an overview of the department's data collection systems, by survey and by administrative sources, with emphasis on the racial, ethnic, socioeconomic position (SEP), and acculturation and language data collected as part of these systems. We identify gaps in the collection of data for measuring health and health care disparities, and conclude that more could be done to effectively capture, measure, and utilize a broader range of federal health data to understand disparities.

We briefly described in Chapter 2 the DHHS initiatives that promote the collection of racial and ethnic data in the department.[1] The DHHS Inclusion Policy requires that information on race and ethnicity be collected in all DHHS-funded and -sponsored data collection systems (both surveys and administrative data systems) and that the latest (1997) OMB standards be used in the collection of these data. In 1999, DHHS released a report entitled *Improving the Collection and Use of Racial and Ethnic Data in Health and Human Services* (U.S. DHHS, 1999), which contains a number

---

[1]It should be noted that these initiatives are focused on the collection of racial and ethnic data and not the collection of SEP or language data. For example, the Inclusion Policy requires the collection of data on race and ethnicity and only encourages the collection of socioeconomic or cultural background characteristics.

of recommendations for the department's data collection programs. This chapter draws upon that work.

HOUSEHOLD AND INDIVIDUAL SURVEY DATA COLLECTIONS

DHHS conducts a number of household surveys that collect information on health and health status, health care utilization, and health care treatment of individuals. The major household surveys and some of their basic characteristics are listed in Appendix A (pages 129-144). These surveys each have different purposes and unique features to address specific questions. The flagship household surveys conducted by DHHS are the National Health Interview Survey (NHIS), the National Health and Nutrition Examination Survey (NHANES), and the Medical Expenditure Panel Survey (MEPS). Each is designed to yield data that are representative of the U.S. civilian noninstitutionalized population. Each survey also collects a broad array of data about health and has special content and design features.

The NHIS is the largest of the surveys and the broadest in data content. It is a continuing survey conducted throughout the year to monitor the health of the U.S. civilian noninstitutionalized population. Approximately 43,000 households comprising almost 106,000 individuals are interviewed each year. The NHIS respondent sample now serves as the sampling frame for the MEPS and the National Survey of Family Growth (NSFG), enabling linkage of the data collected from these three surveys.

The NHANES collects extensive information on health and diet, including a dietary recall of foods consumed by respondents, and includes a medical examination for respondents. The survey is not as large as the NHIS and recently has been conducted less frequently.[2]

The MEPS focuses on health care use, expenditures, sources of payment, health insurance coverage, and health status. It collects data longitudinally from households,[3] interviewing respondents multiple times over a 2-year period.

Other household surveys conducted or developed by DHHS include the Medicare Current Beneficiary Survey (MCBS),[4] the Consumer Assessment

---

[2]NHANES I covered 1971-1975; NHANES II covered 1976-1980; NHANES III covered 1988-1994; and the current NHANES covers 1999-2003.

[3]The MEPS also contains a component on health insurance collected from employers, unions, and other sources of private health insurance, a medical provider component, and a nursing home component.

[4]The MCBS, which surveys 12,000 respondents annually, is a survey of current Medicare beneficiaries that focuses on the financing of health care, but also collects demographic characteristics, health status, insurance status, and information on institutionalization and living arrangements. Racial and ethnic data are collected in this survey. We will discuss this survey below when Medicare data are discussed.

of Health Plans Survey (CAHPS),[5] the National Immunization Survey (NIS), the National Survey of Family Growth (NSFG), the National Maternal and Infant Health Survey (NMIHS), the National Mortality Followback Survey, and the Youth Risk Behavior Surveillance System. In addition, the Behavioral Risk Factor Surveillance System (BRFSS) is a state-level survey developed by DHHS in collaboration with the states to monitor state-level prevalence of behavioral risks among adults (such as drug and alcohol use, presence of diseases such as diabetes, and level of exercise, among other things). The survey contains a core survey that is common across all states so that state comparisons can be made, and states can add their own questions to address the needs of their own populations.

Racial and Ethnic Data Collection

Each of the surveys described above is required to collect racial and ethnic data based on the revised OMB standards. Thus, each at a minimum collects data on whether the individual is white, black, Asian, American Indian or Alaska Native, or Native Hawaiian or Other Pacific Islander, allows respondents to check multiple categories of race, and includes a question on Hispanic ethnic origin. For the BRFSS, racial and ethnic data are collected as part of the core survey and the OMB standards are used. Other surveys collect additional categories of race and ethnicity. For example, the NHIS allows respondents to indicate whether they are Asian Indian, Chinese, Filipino, Japanese, Korean, or Vietnamese, and the National Immunization Survey (NIS) allows respondents to identify themselves as Mexican American, Chicano, Puerto Rican, Cuban, or other Spanish ethnicity.

Socioeconomic Position Data Collection

National household surveys provide some of the most extensive data on socioeconomic position of all the DHHS health-related data systems, with the NHIS, NHANES, and MEPS collecting the most data on SEP.[6] All three of these surveys collect information on employment status, occupation, sources of income and amounts of income from each source, and education levels. Only the MEPS collects information on wealth, requesting

---

[5]CAHPS is a survey tool kit developed by DHHS to survey consumers and purchasers of health plans. A CAHPS-based Medicare questionnaire was developed and has each year, since 1998, been sent to a sample of Medicare managed care enrollees.

[6]See the paper by O'Campo and Burke in Appendix C, for a complete account of which SEP data are collected by DHHS surveys.

the estimated value of different types of assets and debts (e.g., home, business, stock funds, and savings accounts). Although the NHANES and NHIS do not ask about wealth, they both ask about home ownership. The other national household surveys collect limited SEP data—usually only education level or sometimes employment status. The BRFSS core questionnaire asks four questions on socioeconomic position—highest level of education achieved, employment status, health insurance coverage status, and a categorical variable for household income. Many of these surveys do collect information on a household's participation in publicly funded programs for low-income persons. These include Medicaid, SCHIP, and WIC participation. These measures can be used as proxies for low income status.

Acculturation and Language Use Data Collection

Very little data on language use or acculturation are collected in these household surveys. The NHIS collects information on the respondent's place of birth and citizenship. Both the MEPS and NIS collect information on the language in which the interview took place. CAHPS collects information on language spoken at home, and the National Mortality Followback Survey contains information on country of origin.

Data Gaps in National Household Surveys

While the national household surveys that are focused on health issues collect a wide range of data useful for measuring and understanding disparities in health and health care, there are limitations. One major drawback is that because these surveys are designed to be representative of the U.S. population as a whole, and although they generally have large sample sizes, the sample sizes are not usually sufficient to provide statistically reliable estimates of health and health care information for smaller ethnic and racial groups. Sample sizes for some broad racial and ethnic categories (e.g., blacks and Hispanics) are ample in most of these surveys. For example, since 1995, the NHIS has oversampled black and Hispanic populations; and since the MEPS and NSFG both use the NHIS sampling frame, each of these surveys has sufficient sample sizes for improved analysis of these groups. The sample size of the MEPS increased over 50 percent between 2000 and 2002. The 2002 version also oversampled Asians and low-income populations (U.S. DHHS, 2003a). The NHANES currently oversamples black and Mexican American populations, but sample sizes for smaller ethnic and racial groups are often not sufficient to support reliable estimates.

Another weakness of these surveys is that none has a sample size large enough to be representative of all, or even most, of the individual states.

Some policy analysis functions could be served by state-level data on health and health care. For example, it could be useful to compare health care access disparities in states with different policies for providing health insurance for low-income children. As the largest of the health-related household surveys, the NHIS is big enough for reliable estimates of health measures in some larger states and is designed to be readily implemented for use in state-level health surveys, which therefore could be comparable across states and with federal surveys. However, it is currently left to the states to develop and fund any such survey. The BRFSS uniformly collects data on risk behaviors and health practices for all 50 states, and so some state-related policy analyses can be conducted through this survey, but the BRFSS does not collect extensive data on health care, although it does ask about receipt of specific preventive services and about some access problems.

As stated above, the MEPS collects the most extensive data on SEP, including information on wealth; and because it is a longitudinal survey, it collects these data over a period of time, albeit a short period (2 years). Thus tracking changes in SEP and relating them to health outcomes in the short term would be possible with the MEPS, but only on a very limited basis. The other household surveys do not collect extensive data on wealth. There are thus two major limitations of these surveys with regard to SEP data collection: they collect little information on wealth and on measures of income over an individual's life course.

Other federally sponsored surveys with more limited scopes and population coverage do collect extensive data on income, wealth, and socioeconomic position as well as measures of health status. The Health and Retirement Study (HRS), sponsored by the National Institute of Aging and conducted by the University of Michigan, is a panel survey of several birth cohorts all over the age of 50. The survey collects data on health status and income, wealth, and assets, among other items, for over 22,000 individuals each year. Data are from the same household every 2 years. The Assets and Health Dynamics Among the Oldest of the Old (AHEAD) survey, also sponsored by the National Institute of Aging and conducted by the University of Michigan, is a panel survey of individuals either at least 70 years old who responded to the HRS or at least 80 years old and drawn from a sample of Medicare beneficiaries. This survey collects some data on mental, physical, and cognitive health as well as economic, family, and program resources. Both of these surveys collect a rich set of health status and SEP data. Each also oversamples Hispanics and blacks. They do not collect data on health care utilization or treatment and do not cover younger populations. The HRS-AHEAD data have been matched with Medicare claims data to enable researchers to examine relationships between health care treatment and income, wealth, and other demographic background factors. However, sample sizes are too small for many subpopulation analyses, and they may also be too small to study some specific health care problems or treatments. Also, as noted above, these surveys collect very little information on language, nativity, and acculturation.

Although all of the DHHS data sets are required to report race and ethnicity in a standardized way using the new OMB standards, trend analysis over the periods prior to and since implementation of the new standards could create a problem for racial classification. For example, under the old standards individuals of mixed racial backgrounds were asked to choose a single racial background, whereas now under the new standards they can choose multiple racial backgrounds. "Bridging" attempts to statistically model how individuals would respond to the new racial categories based on their responses to the old categories. The OMB has provided guidance on tabulation methods to bridge race responses between the old and new standards (OMB, 2000b). Background analysis for this guidance was conducted using the NHIS.[7] Individuals' responses to race questions from the old and new standards were compared with statistically predicted responses under the old categories based on the individual responses to the new categories. The analysts found that the smallest numerical racial categories were most sensitive to different bridging methodologies.

PROVIDER-BASED SURVEYS

The DHHS sponsors a number of surveys that collect data from hospitals, physicians' offices, and clinics. Some of these surveys collect information directly from the individuals who use these services, but all of them also collect data from the records prepared in conjunction with the service provided. Some major examples of this type of survey are given in Appendix A (pages 144-149). The surveys collect extensive data on the health care utilization and treatment of individuals as well as on the agencies, hospitals, and clinics that provide the care. The National Ambulatory Medical Care Survey (NAMCS), the National Hospital Discharge Survey (NHDS), and the Healthcare Cost and Utilization Project (HCUP) have large sample sizes.

Racial and Ethnic Data Collection

Data on race and ethnicity in these provider-based surveys usually come from records rather than from direct interviews of individuals and are usually recorded by medical personnel or intake workers. Sometimes the

---

[7]Analysis was also conducted with the Current Population Survey and the Washington State Population Survey.

information is based on the observation of the person filling out the record, and sometimes the patient is asked about his or her race and ethnicity. The quality and consistency of the reporting is therefore open to question.

Often, information on race and ethnicity in these surveys is simply missing. For the NHDS, 20 percent of the records do not include data on race and 75 percent do not include data on ethnicity. Similarly, 20 percent of the National Home and Hospice Care Survey records do not include race and 30-40 percent do not include ethnicity. As part of a study to redesign the Drug Abuse Warning Network (DAWN) data collection system in 2002, the Substance Abuse and Mental Health Services Administration (SAMHSA) sent trained researchers into six hospitals' emergency departments to examine their records and abstracts of patients in order to assess which data elements were captured in the records (SAMHSA, 2002). The study found that race and ethnicity were sometimes listed in clinical notes but that the data were not consistently collected. Forty percent of the records lacked information on race and 87 percent provided no information on ethnicity. The study also found that it was unclear whether racial and ethnic categories were consistently used.

Socioeconomic Position Data Collection

Most of these surveys collect only limited data on SEP. In general, only the information needed to ensure payment for the service provided, whether to the individual or to the appropriate government program, is available. In each case, the source of the payment is the only SEP data collected.

Acculturation and Language Use Data Collection

These facilities-based surveys do not collect any data on acculturation and language. Only the National Survey of Substance Abuse Treatment Services collects information on the languages offered for treatment services at the facility.

MEDICARE DATA

Medicare program data have been widely used to study health and health care treatment outcomes, including in studies to measure and understand racial and ethnic disparities in health and health care (Escalante et al., 2002; Escarce et al., 1993; Gornick et al., 1996; Schneider, Zaslavsky, and Epstein, 2002; Skinner et al., 2003). Because of Medicare's entitlement program status and because basic benefits are available to everyone of a certain age regardless of their economic and social background, data from the program have been valuable in better understanding racial and ethnic disparities that exist beyond at least a basic level of health insurance coverage. This section describes the Medicare data system—specifically those data used to measure and understand racial and ethnic disparities in health and health care. The section focuses on the Medicare Enrollment Database (EDB), which is the primary source of racial and ethnic data for linkage with other Medicare records. Previously mentioned surveys such as the MCBS and CAHPS-Medicare Satisfaction Survey also collect data on Medicare enrollees—the MCBS on a sample of about 12,000 enrollees each year and CAHPS on a sample of Medicare managed care enrollees. These two surveys ask questions about race and ethnicity, SEP, and language and acculturation, and thus do not rely on administrative records.

The Medicare EDB contains information on all Medicare beneficiaries. Although it is based entirely on administrative records and does not contain much detailed information on beneficiaries, it is an important database because it can be linked to other Medicare files that include information on health status, service expenditures and financing, age, and gender. Data on race and ethnicity (and on other information about beneficiaries) are obtained from the Social Security Administration (SSA). The SSA provides to the Centers for Medicare and Medicaid Services (CMS), which administer the Medicare program, data on people eligible for Medicare. This information, which is used by CMS to determine who becomes eligible for Medicare, is then used in the EDB once an eligible individual enrolls in Medicaid. Data on race and ethnicity are thus obtained and included in the EDB. However, as we will explain below, racial and ethnic data are not available for all Medicare enrollees and the categories of these data are limited and have changed over the course of the program's history.

Since the beginning of the Social Security program in 1936, racial data were collected on a voluntary basis when a person applied for a social security number (SSN) on the SS-5 form (see Scott, 1999, for a detailed account of racial and ethnic data collection for the Social Security program). The SS-5 form has been the primary source of racial and ethnic data for original and new social security applicants. In 1989, SSA began its "enumeration at birth" program, which assigns an SSN to infants at birth. This system is based on the vital statistics birth registration system. However, information on race and ethnicity from birth certificates is not transmitted to the SSA because it is listed on the birth certificate as "Information for Medical and Health Use Only," meaning it is considered unnecessary for the administration of Social Security programs (Scott, 1999). Thus, for registrants since 1989, no racial and ethnic data are available from SSA unless an individual has applied for a new SSN or a name change, at which time the information was collected.

Over the time for which racial and ethnic data have been collected, the categories of race and ethnicity in the SSA data have changed. Until 1980, the categories collected were white, black, and other; unknown was used to classify those who did not report any race. Since 1980, the categories are white non-Hispanic; black non-Hispanic; Hispanic; North American Indian or Alaska Native; and Asian, Asian American, or Pacific Islander. These data were scheduled to be in compliance with the new OMB standards by the end of 2003. However, data in the 1980 expanded categories were obtained only from people who filled out the SS-5 form (in order to get a new SSN or to request a name change). Data in the new OMB standards will also be collected only when people fill out the SS-5 form. Thus, for people born before 1980 who did not apply for a new SSN, the racial and ethnic categories in the Medicare EDB are still white, black, other, or unknown (Lauderdale and Goldberg, 1996).

In 1994, racial and ethnic data from the new or changed SS-5 records were integrated into the EDB records to correct and fill in missing information.[8] This effort resulted in changes in coding for more than 2.5 million enrollees, about 30 percent of whom were reclassified according to the new racial and ethnic categories implemented after 1980 (Lauderdale and Goldberg, 1996). This update was repeated in 1997 and again in 2000 and 2001 for beneficiaries added since the previous update. CMS's target is to conduct this update for new beneficiaries annually.

CMS also attempted to fill in missing data on race and ethnicity in 1997, using a postcard survey of people with Hispanic surnames, with a Hispanic country of birth (as defined by SSA), and with "other" or "missing" race codes. Over two million people were surveyed but a response rate of only 43 percent was achieved. This effort was nonetheless successful in filling in missing information and resulted in a reclassification of other data for a total of 850,000 people (Arday et al., 2000). CMS has also used beneficiary-level information on race from 32 states for Medicaid enrollees and collected racial information from the End-Stage Renal Disease Medical Evidence Report.

The EDB data on race and ethnicity have been matched and compared with the MCBS data on race and ethnicity for MCBS survey respondents, which are obtained in face-to-face interviews (Arday et al., 2000). Responses to questions about race and ethnicity from the MCBS rounds 1 (1991), 16 (1996), and 19 (1997) were compared with EDB race and ethnicity data.[9] Arday and colleagues found high levels of misclassification

---

[8]Through this process, the wage earner's race is obtained from the SSA records, but the race or ethnicity of any beneficiaries entitled to Medicare through the wage earner is not obtained; rather, the race of the wage earner is recorded as the race of the beneficiary.

[9]In 1991, EDB had only three race categories—black, white, and other. In 1996 and 1997, EDB included those categories plus Hispanic, Asian/Pacific Islander, and American Indian/Alaska Native. The link of the MCBS and EDB in 1997 was after the EDB files were updated with information from new Medicare beneficiaries added since 1994.

of racial and ethnic data in the EDB for groups other than blacks and whites. The sensitivities (or the probability that the EDB correctly classified persons of the given race or ethnicity) were high for the white and black classifications but low for Hispanics, Asian/Pacific Islanders, American Indians, and individuals of other races. The specificities (or the probability that the EDB did not identify someone not of the given race or ethnicity) were high for all nonwhite groups (over 99 percent) but somewhat low for the white classification (87 percent), meaning that 87 percent of those who were nonwhite were not identified as white in the EDB. The sensitivity and specificity of classifications for the other groups improved substantially after the 1994 EDB update; the sensitivity for Hispanics doubled from 19 percent to 39 percent, almost tripled for Asian/Pacific Islanders from 20 percent to 58 percent, and dramatically increased for American Indians from less than 1 percent to 11 percent. Thus, the CMS efforts to fill in missing racial and ethnic data for enrollees had some success in improving the data files. However, misclassification was still high even after the update for these groups, and the authors cautioned against the use of racial and ethnic categories other than black and white in measuring disparities (Arday et al., 2000).

A further limitation in the racial and ethnic data contained in Medicare beneficiary files is that when CMS obtains the enrollee information from the SSA master beneficiary record, it receives information only on the retiree, not the retiree's spouse. Instead, the race of the beneficiary is simply assigned to the spouse.

The EDB does not include any SEP information. The MCBS does collect data on education level and total household income. One possible linkage that would provide a measure of SEP—specifically, current and lifetime earnings income—would be to merge SSA earnings data with the EDB, although SSA earnings records are not perfect measures of lifetime income or of the more general concept of SEP (see Dynan, Skinner, and Zeldes, 2004, for a discussion of the use of SSA earnings as a measure of lifetime earnings). These records are available only for those who have worked in jobs covered by social security, and they do not include undocumented earnings, which could make a sizable difference in earnings measures for immigrant groups (see, for example, Gustman and Steinmeier, 2000). Furthermore, SSA records only cover periods when a person worked and therefore may not be good measures of lifetime income for individuals who did not work their entire lifetime in the United States. The distinction between income and wealth is also important here. Many people in the SSA file who do not have high earnings may have significant sources of wealth. For example, spouses who worked very little may show low earnings, although they may have access to significant resources through their working spouses. Divorced spouses who worked little outside the home while married but who worked later may have had access to higher levels of income during their marriage than what their SSA records imply. Finally, Social Security has a maximum on earnings for which contributions to the system are made. For individuals who meet this maximum, earnings data report only the maximum, not the actual amount of earnings.

Despite such weaknesses in the SSA earnings data, they are a potentially very useful source of SEP data for supplementing Medicare enrollee data. These data are already collected and thus could supplement Medicare enrollee data cheaply relative to the costs of new data collection. The breadth of these records in covering the span of the U.S. working population over the life course (with the exceptions noted above) is unique among available sources of information relevant to studying disparities.

No language data are contained in the EDB. The SS-5 form does collect data on country of origin, but currently those data are not obtained from SSA. The MCBS does not collect information on language or acculturation.

DISEASE SURVEILLANCE SYSTEM DATA COLLECTION

The DHHS has a wide array of data collection systems designed to monitor disease outbreaks, disease treatment outcomes, injuries, food safety problems, and other public health problems. For example, the Haemophilus Influenzae Surveillance System compiles information on all Haemophilus influenzae cases reported to the CDC; the Adult Spectrum of Disease data collection system enumerates and characterizes persons with HIV at various stages of immunologic function; the Firearm Injury Surveillance Study collects information on nonfatal firearm injuries; and the Childhood Blood Lead Surveillance collects information from laboratories on children under the age of 6 who have been tested for blood lead levels. The National Cancer Institute runs the Surveillance, Epidemiology, and End Results (SEER) program to provide data on cancer incidence and survival in the United States. Data are collected from cancer registries in 14 geographical areas covering approximately 26 percent of the U.S. population.[10] Appendix A (pages 149-173) lists these data collection systems, their purposes, and information about the data they collect on race, ethnicity, SEP, and language.

Most of these data collections come from medical records of patient treatments or from laboratories that test for specific diseases. In most cases, an "event" occurs when a person with a disease seeks medical attention and a record of that visit is created. The surveillance data collection systems draw on the medical records to collect information recorded from that

---

[10]We will discuss these state and local cancer registries in more detail in Chapter 5.

initial event and, as applicable, from subsequent visits. Some surveillance data are supplemented with other survey samples of those with the disease, using data gathered at a state or local public health agency that are sent to the federal government. For some systems, only a limited number of states or localities participate in the system so that national coverage of a disease or public health problem is not always possible. Finally, since the data predominantly originate from medical records, they represent people who have a disease or injury and who seek treatment of some sort, not broad demographic populations as are captured in the national health surveys.

Racial and Ethnic Data Collection

The collection of racial and ethnic data in disease surveillance systems is inconsistent. Although racial and ethnic data are collected in many systems, they are often of suspect quality and may not adhere to the OMB standards for such data collection. In many medical record systems, the patient's race is recorded by a health care worker. For example, the HIV/AIDS Reporting System collects racial and ethnic data from a standard CDC form filled out by the provider. The individual with HIV may not be asked his or her race or ethnicity; rather, it may be inferred by the medical staffer filling out the form.

SEER does not use separate questions about race and Hispanic ethnicity; instead, its categories are white non-Hispanic, white Hispanic, black, Chinese, Japanese, Filipino, American Indian/Eskimo/Aleutian, Hawaiian, other, or unknown. Some disease surveillance systems do not collect any racial and ethnic data—for example, the Sexually Transmitted Disease Surveillance System.

Socioeconomic Position and Acculturation and Language Data

SEP data are even more rarely collected in these systems. Education level is collected as part of the Hemophilia Surveillance System. Occupation is collected as part of the Surveillance for Tuberculosis Infection in Health Care Workers system, but this data system is limited to those who work in health care settings. Otherwise, SEP data are not included as part of these data systems and none of the systems collects information on acculturation or language.

HUMAN SERVICES PROGRAMS

DHHS administers several large programs aimed at providing support for poor families and abused or neglected children, child care, early childhood education, and community social services. These programs include Temporary Assistance for Needy Families (TANF), which provides cash assistance and other services to poor mothers with children, the Head Start program, the Child Care and Development Block Grant program, the Social Services Block Grant program, and programs providing grants to agencies that serve abused or neglected children and people abused by family members. Appendix A (pages 173-178) gives background on these programs.

For TANF and Head Start, individuals must apply and meet eligibility requirements to participate in the programs. Information about their race and ethnicity (and in the case of Head Start, their parents' race and ethnicity) is collected as part of the application process. In the TANF reporting system, states must provide data on a quarterly basis to the federal government about the race and ethnicity of persons served. Data on employment, earnings, and income from other sources are also collected. Some states have asset tests and vehicle value asset tests for eligibility, and so these data are also sometimes available. The Social Services Block Grant does not collect racial and ethnic data of persons served through the program. The Child Abuse and Neglect Data system reports data on race and ethnicity.

INDIAN HEALTH SERVICE DATA

The Indian Health Service (IHS) is responsible for providing health care services to American Indians and Alaska Natives. This DHHS agency provides health services—either at IHS facilities or by contract with private-sector providers, tribally operated programs, and urban Indian health programs—to individuals who are members of or can prove descendence from a member of a federally recognized tribe.

As part of its mission and record-keeping functions, IHS obtains data on the utilization of these services. The IHS Patient Registration System collects demographic data on persons that access the IHS system and these data are linked to the IHS patient care information systems. The IHS Ambulatory Patient Care System collects diagnostic data on individuals who receive ambulatory medical care that is either provided or funded by IHS. The IHS Dental Services Reporting System and Inpatient Care System serve the same function for these health service providers. IHS also maintains birth and death records for American Indian and Alaska Native individuals. These records are forwarded to NCHS from the states and then forwarded to IHS.

The data in these systems are used both to monitor health status (e.g., infant mortality, life expectancy) and to understand health care utilization and treatment of American Indian and Alaska Native populations. IHS produces a series of reports called *Trends in Indian Health and Regional Differences in Indian Health*, which uses these data. None of these data systems include information on SEP or language.

RECOMMENDATIONS

There are many excellent sources of data collected by DHHS that can be utilized to better understand disparities in health and health care. But, as noted throughout this chapter, limitations in these data sources exist. Through its Inclusion Policy and its 1999 report *Improving the Collection and Use of Racial and Ethnic Data in Health and Human Services*, which included recommendations to improve its racial and ethnic data collection, the department has begun to address some of the data weaknesses highlighted in this chapter. In this section, the panel gives its recommendations for improvements to national data collection efforts.

The 1999 DHHS report is a very comprehensive presentation of the federal issues related to racial and ethnic data. The reports' recommendations are important for improving federally based data sources and should be acted on. The report calls for the authoring groups to develop a plan to implement the recommendations that would prioritize recommendations, create a detailed plan of action, establish a responsible office(s) to carry out the plan, and consider costs needed for implementation. Thus far, no implementation plan has been produced, although some DHHS agencies have implemented some of the recommendations. The panel believes that DHHS should develop such a plan and continue to implement the data improvement recommendations. The plan should include the establishment of a body that would be responsible for coordinating implementation across the various agencies of the department and for ensuring that agencies follow through with recommendations.

RECOMMENDATION 4-1: DHHS should begin immediately to implement the recommendations contained in its 1999 report entitled Improving the Collection and Use of Racial and Ethnic Data in Health and Human Services.

There are many important recommendations in the 1999 DHHS report. The panel wishes to emphasize a few that it sees as priorities for the department and vital to the improvement of federal data collection systems. The panel's primary focus is on the improvement of existing data collection efforts to make them more effective and the creation of new collections to fill data gaps.

The panel believes that four themes in the 1999 DHHS report are especially noteworthy:

(1) developing feasible approaches for including racial and ethnic groups in national surveys;
(2) improving the collection and analysis of SEP, language, and acculturation data;

(3) ensuring the collection of racial and ethnic data in DHHS and DHHS-sponsored administrative record systems; and

(4) developing mechanisms for linking records across government data systems.

National household surveys are not large enough to support analysis of health outcomes for many racial and ethnic subgroups. The costs of obtaining extensive health data—such as the data collected in surveys like the NHIS or the NHANES—for small or geographically concentrated racial and ethnic groups make it impossible to collect such data on a regular basis for every racial and ethnic group. However, periodic studies targeted to survey specific groups in specific areas could be conducted and could provide vital data on the health outcomes of these groups. The panel therefore recommends that DHHS develop a schedule for special surveys of population subgroups—e.g., American Indians in Washington state, Cuban Americans in Florida. These are just two examples of groups that could be surveyed. In developing the plan, DHHS would need to identify specific information needs in consultation with the various DHHS agencies and representatives of subgroups. Such targeted surveys may also consider adding appropriate, more extensive measures of acculturation. This schedule, covering a 10- to 20-year period and each year identifying the group to be targeted, would be a feasible way of collecting meaningful data on racial and ethnic subgroups over time.

RECOMMENDATION 4-2: DHHS should conduct the necessary methodological research, and develop and implement a long-range plan, for the national surveys to periodically conduct targeted surveys of racial and ethnic subgroups.

The panel notes that the DHHS Assistant Secretary for Planning and Evaluation (ASPE) has sponsored work to assess federal health data sets for their ability to provide data on detailed Asian and Hispanic subgroups and on American Indian and Alaska Natives (Waksberg, Levine, and Marker, 2000).

Beyond sample size, there may be other statistical issues to address when targeting certain racial and ethnic groups (Kalsbeek, 2003). The rarity of these groups, combined with their geographic dispersion, often makes it inefficient to sample them with greater intensity in household samples involving the usual practice of selecting samples of residential telephone numbers or geopolitical area units, such as counties and census block groups. Moreover, in addition to reducing the likelihood of sample coverage, the relative mobility of some types of racial and ethnic groups (e.g., recent immigrants, farm workers) often leads to skewed samples favoring those who are more mobile, unless specific steps are taken in sampling and estimation to correct these problems.

The issue of comparability in measurements is also important. For example, recent Spanish-speaking immigrants may understand and thus respond to survey questions differently from U.S.-born Hispanic respondents who speak English. Special efforts may therefore be needed to develop survey questions that can be uniformly understood. The potential for loss in comparability in the process of information exchange in general population surveys can mask or exaggerate real differences in studies to assess disparity. For these reasons, DHHS should take steps beyond those currently being taken in its surveys to enhance the department's ability to determine where disparity exists and evaluate attempts to eliminate it.

RECOMMENDATION 4-3: The adequacy of sampling methods aimed at key racial and ethnic groups, as well as the quality of survey measurement obtained from them, should be carefully studied and shortcomings, where found, remedied for all major national DHHS surveys.

The DHHS Inclusion Policy clearly states the goal of collecting racial and ethnic data for all department programs and record collections and surveys. The department's household surveys all collect racial and ethnic data in accordance with OMB standards. However, the department's health data frequently come from administrative records either from DHHS programs (e.g., Medicare) or from clinics, providers, and laboratories. Not all of these records use the OMB standard categories for race and ethnicity, and some do not collect such data at all; as a result, the racial and ethnic data collected through these records are inconsistent and unstandardized.

These administrative data sets contain a large number of records and could, with better data on race, ethnicity, SEP, and language use, offer a valuable source of information for research on disparities. In order to improve these data sets, the department should enforce the Inclusion Policy and require those programs that do not report racial and ethnic data to collect such data in accordance with the OMB standards. This is especially important with respect to the Health Insurance Portability and Accountability Act (HIPAA) standards.[11]

RECOMMENDATION 4-4: DHHS should require the inclusion of race and ethnicity in its data systems in accordance with its Policy for Improving Race and Ethnicity Data.

---

[11]HIPAA establishes a national standard for electronic transactions with which all health plans, health care clearinghouses, and providers conducting business electronically must comply. The law requires that DHHS adopt transaction and code set standards for covered transactions, including claims and enrollment transactions. We will discuss HIPAA and its relevance to the reporting of racial and ethnic data in more detail in Chapter 6.

Although DHHS data systems do not consistently collect data on SEP, such data are needed both to better understand racial and ethnic disparities and to identify effects on deprived groups that are not defined by race or ethnicity but that experience health or health care disparities.

The national household surveys generally provide sufficient SEP data, obtaining some measures of employment, education, insurance coverage, income, and wealth. There are weaknesses, however. Income and wealth data are often not collected in much detail. Income questions in some of these surveys are often categorical and do not obtain information on exact income levels or on how much income is received from different sources. The collection of wealth data is rarer even though wealth is a very important measure of a lifetime accumulation of resources.

The department's administrative and record-based data collections include very little SEP data. In most cases, only information on insurance or method of payment is recorded. Sometimes employment and educational attainment status are collected as well. With the understanding that any data collected in these systems must be relevant to the administration of the program or service, the department should consider ways to collect more SEP data, both in surveys and in administrative data collection.

Knowledge about health and the health care system and the ability to communicate with health care providers are crucial components of an individual's ability to negotiate the health care system, understand diagnoses and recommended treatments, and pay for treatment. Those who are not proficient in the English language or who do not know the system well may have more difficulty getting the care they need. Data on language proficiency and acculturation could be used both to explain differences in health outcomes across and within ethnicities and to improve health care services and programs so that they better accommodate these populations.

Very little information on language use and acculturation is collected in national health surveys and even less is collected in DHHS administrative and surveillance system records. Surveys are best suited to collect more extensive information on language ability and on the degree of acculturation, whereas records-based collections are more limited in the extent of data that can be obtained. However, in many instances, information on primary language could be useful both for the provision of medical services and information and as a data element for later use in research.

RECOMMENDATION 4-5: DHHS should routinely collect measures of SEP and, where feasible, measures of acculturation and language use.

Weaknesses in a single source of data can often be remedied by linking data from other sources, a practice that can make use of existing data without the burden of new data collection. But there are sometimes barriers to linking across agencies and even within agencies. For example, the ability to match data may be limited if common identifiers are unavailable or of poor quality. Confidentiality concerns also arise with data linkages because a common identifier is needed in both data sets to link the data and this may increase the possibility that an individual's identity can be recognized. Special attention must therefore be devoted to the protection of respondent confidentiality and proper use of the data with linked data sets.

Although there are barriers and costs to sharing data, the resulting richer sets of data can be used to fill in important gaps in any single data source. For example, matching SSA earnings records to Medicare claims data provides a means to understand links between race, ethnicity, and SEP and health care treatment and treatment outcomes. Therefore, where possible, the department should encourage and promote data linkages, including between data sets collected and maintained both in different DHHS agencies and with non-DHHS departments or institutes.

> **RECOMMENDATION 4-6: DHHS should develop a culture of sharing data both within the department and with other federal agencies, toward understanding and reducing disparities in health and health care.**

Each of these first six recommendations (and the last two in this chapter) is directed to the department in general, and not to a specific agency within DHHS. The panel directs these recommendations to the Office of the Secretary because the panel believes the actions of these recommendations need to be taken on a department-wide basis and because there is no other agency within DHHS to which these recommendations can be directed. As was mentioned above, the 1999 DHHS report calls on DHHS to establish a responsible body for coordinating the implementation of that report. Such a body would be a logical place to direct the panel's recommendations, but it does not yet exist.

Data on Medicare enrollees, which cover all elderly persons who enroll in Medicare and are collected through the EDB, are crucially important for understanding disparities in health and health care treatment. Since much of the health care a person receives occurs later in life, the database covers individuals when they are likely to be using the health care system most frequently.

As this chapter discussed, the reporting of racial and ethnic data in Medicare is incomplete. Many individuals enrolled in Medicare do not have a reported race or ethnicity in their records, and, under current procedures, many who will eventually qualify for Medicare in the future will not have racial and ethnic data in their records. Because of the importance of Medicare data in measuring health and, especially, health care disparities, the panel believes it is crucial for CMS to take the initiative in collecting racial and ethnic data for both current and future enrollees. For new enrollees, the best time to collect data on race and ethnicity, SEP, and language would appear to be at the time of enrollment. A very brief questionnaire could be used for both current and new enrollees. To keep the survey short, complete information about income and wealth need not be collected, although a categorical question on income or educational level could be included. A question about language use might also be considered.

The collection of additional data for current enrollees will not be an easy or inexpensive task. A previous CMS attempt to collect racial and ethnic data through a postcard survey of current enrollees achieved some success in filling in vital missing information (Arday et al., 2000), but it had a poor response rate. A thorough effort is needed with full support for proper follow-up.

> **RECOMMENDATION 4-7:** The Centers for Medicare and Medicaid Services should develop a program to collect racial, ethnic, and socio-economic position data at the time of enrollment and for current enrollees in the Medicare program.

As mentioned, the Medicare Enrollment Database does not collect SEP information. It is possible to obtain records of an enrollee's earnings and employment histories through the wage history files of the SSA, but these data are not without problems. For example, some individuals may not have worked long and may thus show relatively lower earnings in the system. Or, some individuals may have had a spouse who earned wages and thus have a greater income or wealth than the SSA records imply. In addition, those who have worked in the U.S. labor market for only a few years may also show low earnings even though they may have accumulated wealth from income earned outside the SSA system (Gustman and Steinmeier, 2000; Dynan, Skinner, and Zeldes, 2004). These data should therefore be combined with information on the number of quarters the individual was in the system (since reporting is quarterly) and the number of years the individual earned the maximum contribution level. Privacy and confidentiality concerns should also be considered carefully. However, despite these potential barriers, the panel believes that the CMS and SSA should cooperate to link these two important data sets.

> **RECOMMENDATION 4-8:** The Centers for Medicare and Medicaid Services should seek from the Social Security Administration (SSA) a summary of wage data on individuals enrolled in Medicare.

The panel recognizes that there are barriers to obtaining these data; for example, privacy and confidentiality concerns have hindered CMS efforts to obtain such data in the past. As an alternative to obtaining earnings records, CMS is currently seeking to use information on the amount of social security benefits paid to individuals (which are based on earnings) as a proxy for earnings. This information can be obtained from the SSA Master Beneficiary Record, from which CMS has previously obtained data.

Leadership for Implementing OMB Standards for Health Data Collection

There is still a lack of understanding and delay in implementation of the new OMB standards for collecting racial and ethnic data, particularly outside the federal statistical community. Many attendees of the panel's Workshop on Improving Race and Ethnicity Data Collection expressed confusion over what the minimum race categories were, whether more detailed categories could be used, how data on Hispanic ethnicity should be collected, how multiple-race responses should be handled, and how data collected before the new standards can be bridged to data collected since the standards were implemented (National Research Council, 2003).

The OMB has published materials to guide researchers in using the new standards and bridging to the old categories.[12] But while federal researchers may be well aware of these standards, researchers at nonfederal levels may not be. DHHS should take a leadership role in educating relevant staff at all DHHS agencies, state health agencies, and private entities that collect data for DHHS programs about these new standards. The department should increase awareness of the OMB standards by disseminating the appropriate OMB materials to the various state and private entities from which DHHS obtains data. For example, CMS could distribute such materials to all state Medicaid directors. In addition, DHHS should assume responsibility for ensuring that the new standards are properly and consistently applied throughout the department's data collection systems. Because it is often necessary to collect data for more specific racial and ethnic groups than are listed in the OMB standards, the panel also recommends that DHHS promote uniformity in such data collections by publishing suggested subclassifications for each of the OMB classifications for use in all DHHS data collection efforts. Such steps would be mutually beneficial both for states and private entities, who are looking for guidance on the OMB standards, and for DHHS, which relies on states and private entities to provide much of its data on health and especially health care-related topics. As the following chapter will discuss, the data collection efforts of many states are conducted for federal programs or in cooperation with federal agencies to obtain national-level data. DHHS can improve the state-based collections of information on race and ethnicity by being consistent across the department in requiring OMB standards in its data collection programs. Inasmuch

---

[12]See for example, OMB (2000a and 2000b).

as some of these cooperative data collection efforts require input and cooperation from each of the states, implementing the standards in states may be a delicate balance of allowing states to meet their own needs for data collection while promoting national-level comparability.

RECOMMENDATION 4-9: DHHS should prepare and disseminate implementation guidelines for the Office of Management and Budget (OMB) standards for collecting racial and ethnic data.

The panel notes that some agencies within DHHS have issued such guidelines. For example, the National Institutes of Health have issued guidelines for maintaining, collecting, and reporting racial and ethnic data in clinical research.

The Importance of SEP in Understanding Racial and Ethnic Health Disparities

In Chapters 2 and 3, we illustrated the interrelationships between race, ethnicity, and SEP. Because of the interrelationship of these variables, in order to accurately interpret racial and ethnic differences in health and health care it is important to consider differences within groups of different social and economic backgrounds. Therefore, where possible, the panel urges DHHS to report health and health care disparities across different levels of SEP. The specific SEP measures used may depend on the outcome of interest (for example, education level may be the most appropriate measure for examining preventive health knowledge and outcomes—such as the percent of women receiving a mammogram each year) or upon the availability of data. In any case, the department should make an effort to consider SEP differences in conjunction with racial and ethnic differences in its health disparities reports.

RECOMMENDATION 4-10: DHHS should, in its reports on health and health care, tabulate data on race and ethnicity classified across different levels of socioeconomic position.

5

State-Based Collection of Data on Race, Ethnicity, Socioeconomic Position, and Acculturation and Language Use

State health agencies collect a large amount of data on health and health care services to aid in their missions of providing health programs and services to the populations of their states. States also collect these data as part of their missions to license and regulate health care providers and insurers, train and distribute a health workforce, and to generate measures for market and policy decisions. In addition, they collect such data to meet federal data collection requirements for programs that are run by individual states with some funding from federal sources.

The federal government relies on states for much data collection because its regulatory powers and service provision activities are not nearly so broad. The states, in fact, are the collectors of much of the data used to study health and health services at both the state and federal levels. This is why there are many cooperative data sharing efforts, such as the Vital Statistics Cooperative Program (VSCP) for vital statistics, the Healthcare Cost and Utilization Program (HCUP) for hospital discharge data, the Surveillance, Epidemiology, and End Results (SEER) program for cancer, the Behavioral Risk Factor Surveillance System (BRFSS), and Medicaid. In this chapter, we briefly discuss the major state-based data collection systems and the racial and ethnic, socioeconomic position (SEP), and acculturation and language data that are collected in them. The systems reviewed are vital statistics birth and death records, hospital discharge abstracts, cancer registries, health interview surveys, and Medicaid and the State Children's Health Insurance Program (SCHIP), with examples of how these data have been used to understand disparities in health and health care. We also discuss gaps in these state-based data collection systems. The chapter concludes with recommendations calling for states to push for the collection of data on race, ethnicity, SEP, and acculturation and language use as much as possible, and for the Department of Health and Human Services (DHHS) to provide technical assistance, resources, and incentives to states to improve the collection and use of these data.

The panel commissioned a background paper on state-based data collection for its Workshop on Racial and Ethnic Data in Health (see the paper by Geppert et al., in Appendix E). This paper, which was presented at the workshop, drew on interviews with officials from four states about their racial and ethnic data collection as case studies on how states use such data and the problems they encounter in collecting and analyzing them. This chapter draws on the results presented in the paper.

In considering state-based data collection, it is important to note that states' needs for social and demographic data on their populations are different from those of the federal government. State governments directly implement and evaluate health intervention programs in order to ensure the health of their populations. While the federal government also implements health programs and monitors public health, it does so, for the most part, in an indirect way through contracts that is more removed from actual implementation of the programs (the Indian Health Service is an exception).

The federal government must monitor the nation as a whole and cannot, because of resource constraints, regularly collect data on ethnic groups that do not represent a sizable portion of the national population. It therefore aggregates data on racial and ethnic groups into generally broad categories. States, on the other hand, must provide services for their own populations, which may include concentrated populations of particular ethnic groups; for example, Dominicans and Puerto Ricans in New York, Cubans in Florida, Hmong in Minnesota, and Mexicans in Texas.

These subgroups may have differing health and health care needs. For example, Puerto Ricans are more likely than other Hispanics to have low-birthweight babies (9.3 percent of live births compared with the 6.5 percent national average for all Hispanics) (National Center for Health Statistics, 2003); and Hawaiians have a lower rate of early prenatal care than other Asian and Pacific Islanders (79.1 percent, compared with 90.1 percent for mothers of Japanese descent, 87.0 percent for mothers of Chinese descent, 85.0 percent for mothers of Filipino descent, and 82.7 percent for mothers of other Asian or Pacific Islander descent) (National Center for Health Statistics, 2003).

For these reasons, the broad OMB categories may not be as appropriate for all states as they are for most federal-level data collection. States need data for specific ethnic subpopulations in order to target public health interventions and measure health outcomes and disparities, and these subpopulations may not correspond to the broad racial and ethnic categories stipulated by the federal government. It should be noted, however, that the OMB standards establish minimal racial and ethnic categories and therefore do not prohibit the use of more detailed categories.

VITAL STATISTICS BIRTH AND DEATH RECORDS

Each state issues birth and death certificates as part of the country's vital statistics system. These data provide states with information that can be used to assess and improve the health of the population. For example, states use vital statistics for prenatal care interventions and infant mortality reduction. Information for birth certificates is recorded at the birth of an infant by a health care professional, and information for death certificates is usually collected by a funeral home director. It is believed that these two record systems are essentially complete in their coverage of births and deaths in the United States (see U.S. DHHS, 1997).

All states and territories provide these core data to the National Center for Health Statistics (NCHS) under the Vital Statistics Cooperative Program (VSCP). The program (and its predecessors) was implemented to record national vital statistics and to encourage comparable reporting of these events across states and U.S. territories. Standards for the reporting of minimum basic data items were developed (and continue to be reassessed) by NCHS working with state vital statistics organizations. States are funded to provide the standardized data to NCHS, with each state's federal funding level based on its reporting of these minimum basic data.

Standards for reporting racial and ethnic data are included as are standards for reporting the education levels of the parents (on birth certificates) or of the decedent (on death certificates). For birth certificates, the race and ethnicity of the infant are not reported; rather, the race and ethnicity of the infant's mother and father (if known) are reported. The education levels of both parents are recorded in the same manner, as are their countries of origin. The form is filled out by a medical records clerk, who is supposed to ask the parents for this information. In the case of death registrations, the race and ethnicity and education level of the deceased are usually recorded by the funeral home director or a health care worker who requests the information from either the decedent's next of kin or a family representative while filling out the forms. No data are collected on the decedent's country of origin or language.

The racial and ethnic categories currently used in the vital statistics system were reviewed as part of the regular vital statistics standard certificate review process (NCHS, 2000). This review resulted in a recommendation for expanded racial and ethnic categories that would include separate Asian categories (Asian Indian, Chinese, Filipino, Japanese, Korean, Vietnamese, and other Asian) and separate Pacific Islander categories (Native Hawaiian, Guamanian or Chamorro, Samoan, and other Pacific Islander)—all of which can be aggregated to the five minimum OMB categories—and "specify" lines where individuals can indicate their tribe or "other" status. In addition, the review panel recommended that question about Hispanic ethnicity be listed before the race question. The recommended categories are shown in Box 5-1. The recommendations of the review are still being considered and have not yet been implemented.

Several studies have examined the quality of racial and ethnic data from vital records. Hahn, Mulinare, and Teutsch (1992) used birth records linked with infant-death records to study the consistency of racial and ethnic reporting for infants who died before their first birthday. The study compared the race and ethnicity coded on the infant's birth record (which is determined by the parents' report of race and ethnicity) and the infant's death record (which is usually recorded by observation from a funeral director or other certifier) and how these inconsistencies affected computa-

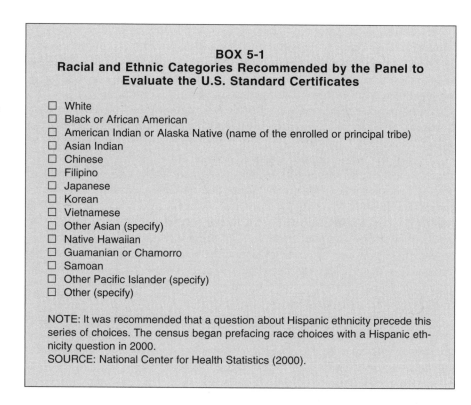

BOX 5-1
Racial and Ethnic Categories Recommended by the Panel to
Evaluate the U.S. Standard Certificates

☐ White
☐ Black or African American
☐ American Indian or Alaska Native (name of the enrolled or principal tribe)
☐ Asian Indian
☐ Chinese
☐ Filipino
☐ Japanese
☐ Korean
☐ Vietnamese
☐ Other Asian (specify)
☐ Native Hawaiian
☐ Guamanian or Chamorro
☐ Samoan
☐ Other Pacific Islander (specify)
☐ Other (specify)

NOTE: It was recommended that a question about Hispanic ethnicity precede this series of choices. The census began prefacing race choices with a Hispanic ethnicity question in 2000.
SOURCE: National Center for Health Statistics (2000).

tions of infant mortality rates (IMRs). Results showed that 3.7 percent of the infants in the study sample were classified as having a different race at death from that recorded at birth. Inconsistencies in classification were more prevalent for nonwhites, and especially for infants of "other" race classifications. More infants were classified as white at death than at birth (i.e., there were fewer infants classified as black at death compared to the number at birth, and fewer infants classified as other race at death than at birth). Finally, the authors found that if consistent racial and ethnic coding is used to calculate IMRs,[1] the IMRs for whites and non-Hispanic whites decrease compared with those for the same groups using standard race definitions. However, the IMRs for blacks, non-Hispanic blacks, and other race groups increase compared to IMRs calculated using standard race definitions. A similarly structured study in the state of Washington found that 61 percent of infants who died in the first year of life and were recorded as being of American Indian or Alaska Native descent on their birth records were coded as American Indian or Alaska Native at death (Frost and Shy, 1980).

Other studies have found high-quality information on race and ethnicity in vital records for most racial and ethnic groups. Baumeister and colleagues (2000) found that racial and ethnic information on birth records was quite similar to the respondent's self-identified racial and ethnic information provided in face-to-face interviews. The one group for which birth records did not match self-reported racial and ethnic categorizations at a high rate was Native Americans. Sorlie, Rogot, and Johnson (1992), using data from the National Longitudinal Mortality Study, showed that classification of blacks and whites in death records is highly comparable to racial and ethnic classifications from survey data reports for blacks and whites. However, the classification of American Indians and Asian and Pacific Islanders on death records had more errors. Individuals from these two groups were often categorized as white on death records. The authors suggested that these misclassifications could result in underestimation of death rates for these groups. Frost and colleagues (1994) came to a similar conclusion regarding data on death certificates for American Indians and Alaska Natives in Washington state. This study found that 12.8 percent of individuals who appeared in the Indian Health Service (IHS) patient registry (that is, patients treated at IHS facilities, who must be a member or descendent of a member of a federally recognized tribe) in Washington state were not classified as American Indian or Alaska Native on their death

---

[1]A different approved rule for assigning race at birth was also implemented to calculate the new IMR.

records. The authors concluded that death rates for this group may be underestimated.

HOSPITAL DISCHARGE ABSTRACTS

Thirty-seven states have legislative mandates to collect discharge abstracts on patients that use nonfederal hospitals. In most states without a mandate, providers may submit the data voluntarily to a private entity. These data contain a complete demographic, treatment, and financial record for each patient admitted to the hospital and are the only source of population-based health care data (Chapter 6 discusses these abstracts in more detail). States use the data to evaluate health care access, cost, and quality and to support policy and market-based decisions. Medicaid programs use the data for comparative norms for uncompensated care estimates. Most states that collect these data use a standard format, the Uniform Bill for Hospitals (UB92), a form that is used to pay claims. Although the core UB92 data elements do not include standard racial and ethnic data, 27 states include these data as part of their own requirements (see the paper by Geppert et al., in Appendix E).

The federal government collects and maintains hospital discharge data as an important source of information to study health care utilization and treatment outcomes. The government purchases the state hospital discharge data from state and private data organizations and hospital associations to create a national data set, the Healthcare Cost and Utilization Project (HCUP) maintained by the Agency for Healthcare Research and Quality (AHRQ). HCUP is a family of databases and products and includes the largest collection of longitudinal hospital care data in the United States, with all payer, discharge-level information since 1988. The AHRQ Quality Indicators provide information about health care cost, quality, and access derived from the discharge data. States can measure inpatient admissions to hospitals for preventable conditions such as diabetes or asthma. If the state captures data on race and ethnicity with its discharge data, admission rates for Hispanics and other minority groups can be measured and used to target interventions. Data from HCUP have also supported disparities research. Elixhauser and colleagues (2002) used data from the Healthcare Cost and Utilization Project, which is described further in Chapter 6, to examine disparities between Hispanic and non-Hispanic whites with cerebrovascular disease in the use of in-hospital diagnostic and therapeutic procedures. The study found that controlling for a hospital's experience with Hispanic patients eliminated or reduced the magnitude of disparities in procedure use.

Because HCUP data are collected at the state level and states vary in their data collection practices, many fields, such as race and ethnicity, are incomplete or inconsistent in their format.

CANCER REGISTRIES

Forty-five states collect data on cancer cases, including information about the occurrence of cancer, the type and location, the conditions of the cancer at diagnosis, and treatment. These data are used to study the causes of cancer and outcomes of treatment, and to target intervention and prevention programs. The registries are sponsored by either the National Cancer Institute's Surveillance, Epidemiology, and End Results (SEER) program or the CDC's National Program of Cancer Registries (NPCR).[2] Data come from hospitals, physicians' offices, and laboratories and are provided to a central statewide registry, or, in the case of six SEER metropolitan sites and other SEER rural or special population sites, to local registries. The data are then shared with either the National Cancer Institute (SEER sites) or the CDC (NPCR sites) or, in some cases, both.

The CDC provides funding and technical assistance to NPCR participant states to collect the data and to improve their completeness, timeliness, and quality. To receive funding, states must implement quality and format standards that are reviewed and certified by the North American Association of Central Cancer Registries (NAACCR). These standards include prescribed formats for the collection of both racial and ethnic data.[3] SEER abstracts racial and ethnic data that are as detailed as are available from the records used to identify individual cases of cancer. For example, if a record shows the individual is of Chinese descent, SEER will include that information in its record for that incidence of cancer. The SEER system also aggregates these racial and ethnic data into one of four broad race categories (black, white, Asian or Pacific Islander, and American Indian or Alaska Native) and into Hispanic or non-Hispanic ethnicity. In 2002, 28 states participated in NPCR and were certified as meeting the data standards requirements, 14 participated but were not certified, 4 states participated in

---

[2]See the Web sites for SEER (http://seer.cancer.gov/) and NPCR (http://www.cdc.gov/cancer/npcr/).

[3]The racial categories used are white, black, American Indian/Aleutian/Eskimo, Chinese, Japanese, Filipino, Hawaiian, Korean, Asian Indian/Pakistani, Vietnamese, Laotian, Hmong, Kampuchean, Thai, Micronesian, Chamorran, Guamanian, Polynesian, Tahitian, Samoan, Tongan, Melenisian, Fiji Islander, New Guinean, other Asian (including Asians, not otherwise specified [NOS] and Oriental), Pacific Islander NOS, other, and unknown. Ethnicity categories are non-Spanish/non-Hispanic, Mexican (includes Chicano), Puerto Rican, Cuban, South or Central American (except Brazil), other Spanish (includes European), Spanish NOS, Hispanic NOS, Spanish (surname only), and unknown whether Spanish or not.

both NPCR and SEER and were certified, and 4 participated in SEER alone.[4]

Cancer registries do not include much information on socioeconomic position. While work history is often contained in these records, usually to understand exposure to environmental factors, this information is often missing or incomplete (Swanson, Schwartz, and Burrows, 1984). However, these systems include geographical information on the patient that would allow for the geocoding of aggregate-level SEP data from the person's record. Singh and colleagues (2003) linked SEER information to county and census tract-level information on poverty rates to study cancer incidence, diagnosis, and mortality across areas with differing poverty levels.

STATE HEALTH INTERVIEW SURVEYS

A number of states and local areas conduct their own surveys of individuals to collect data on health status and health care services. These surveys typically sample the population of the state or local area. State surveys are useful because they can be used for state-level estimates and allow for details on subgroups (by race and ethnicity and/or by geographic area such as county or region). For example, the Hawaii Department of Health has conducted the Hawaii Health Survey every year since 1968. It is a household survey of all persons living in noninstitutionalized housing units in Hawaii and collects demographic and health data to monitor the health, socio-demographic status, and other characteristics of the population of Hawaii. It also provides statistics for planning and evaluation of health services and programs, and identification of problems (see http://www.hawaii.gov/doh/stats/surveys/hhs.html). Based on this survey, the Hawaii Department of Health publishes annual reports on the general demographics, income, health insurance, and health conditions of the state's population (see Hawaii Department of Health, 2003). Many health conditions are reported by ethnicity.

The state of California in 2001 sponsored the California Health Interview Survey, which collected extensive health-related data on individuals—demographic information, health status, health insurance coverage, eligibility and participation in public coverage and assistance programs, utilization of health services, and access to care. These data have been used, for example, to show that racial and ethnic differences exist in the percent of adults with diabetes in California who monitor their glucose levels at least once per day and in percentages for those with different health insurance

---

[4]It is not clear if the racial and ethnic data format requirements were the reason states were not certified.

statuses (Diamant et al., 2003). This study found that American Indians and Alaska Natives were the most likely to monitor their glucose level, while Latinos were the least likely to monitor. Not surprisingly, those without health insurance were also least likely to monitor their glucose, whereas those with Medi-Cal (California's Medicaid program) were the most likely.

King County (Seattle), Washington, conducted a survey modeled on the Behavioral Risk Factor Surveillance System called the Ethnicity and Health Survey, which oversampled seven racial and ethnic minority groups, including five different Asian ethnic groups with high concentrations in the Seattle area (Smyser, Krieger, and Solet, 1999).

Many states use surveys developed on a national level for implementation by states and localities that choose to use them. Examples of these surveys include the Behavioral Risk Factor Surveillance System (BRFSS) (described in Chapter 4) and the Pregnancy Risk Assessment Monitoring System (PRAMS) (described in Chapter 3), which collects information from women with new infants on their behavior before, during, and after pregnancy. Each of these surveys contains a core set of questions that can be supplemented by each state, with additional questions according to its data collection needs.

One model of a federal-state data partnership is the Health Resources and Services Administration's (HRSA) State Planning Grants. HRSA provided funding to states to develop plans for providing access to affordable health insurance coverage to all citizens. In 2002, 32 states received funding to collect data for and analysis of the characteristics of the uninsured and to develop potential models to increase coverage.

MEDICAID AND SCHIP DATA

Medicaid provides health insurance coverage for low-income families with children, disabled individuals, and certain low-income elderly citizens. It is a program with shared state and federal funding that is coordinated by the Centers for Medicare and Medicaid Services (CMS) in DHHS. SCHIP is a federally funded, state-administered program that provides insurance coverage for low-income children from birth to age 18. States can choose to use federal SCHIP funds to expand their Medicaid programs or as separate state health insurance programs. SCHIP served 5.3 million children in 2002.

States collect information on Medicaid program enrollees to assess eligibility and administer the program. Until the Balanced Budget Act of 1997, states were required to report only aggregate data to the federal government, although they could voluntarily report individual-level data on enrollees. When the voluntary data collection ended (1998 was the last year), 32 states had submitted data on individuals (including racial and ethnic data) to the federal government. The 1997 act mandated that states report data on eligibility and claims through the Medicaid Statistical Information System (MSIS) on a quarterly basis beginning in 1999. These data have not yet been extensively used for research purposes.[5]

The MSIS required data set includes information on eligibility (e.g., characteristics of Medicaid enrollees, including racial and ethnic data in addition to financial information) and on claims made on their behalf (e.g., utilization of health care services and payments). States have the discretion to collect racial and ethnic information as they see fit; for those that do, CMS, through MSIS, requests that the data be reported in a standardized format. Currently, racial and ethnic data are reported to CMS in a set of categories that combines race and ethnicity, with no separate question for Hispanic ethnicity. The categories of race and ethnicity collected by CMS from the states are: white; black or African American; Asian; Hispanic or Latino (no race information available); Native Hawaiian or other Pacific Islander; Hispanic or Latino and one or more races; more than one race (Hispanic or Latino not indicated); and unknown. States do have the option, however, of reporting race and Hispanic ethnicity in separate questions, indicating whether an enrollee is white or not, black or African American or not, American Indian/Alaska Native or not, Native Hawaiian or Other Pacific Islander or not, and Hispanic or not.[6] Thus, if states record this information separately, information about Hispanic ethnicity can be obtained separately from racial status in the MSIS. In addition, if a state collects this information, the MSIS data include an option for multiracial status and so multiple races can be identified as well. To accommodate the additional data collection, however, each state program will have to implement form changes, computer system changes, and staff training.

CMS does not yet have any information on the quality of the racial and ethnic data collected through the MSIS. In fiscal year 2000, the race and ethnicity of 3 million (about 7 percent) of the total 44.5 million Medicaid enrollees were reported as "unknown." Furthermore, CMS does not know how the racial and ethnic data categories collected through each state's eligibility processes are translated into the MSIS categories for race and ethnicity. For example, if a state combines the categories Asian and Pacific Islander in collecting information on enrollees and potential enrollees, it is

---

[5]The DHHS Assistant Secretary for Planning and Evaluation (ASPE) has funded a study to review the potential for state-based data reported under the MSIS to be used to study managed care enrollment, long-term care service use, and mental health service use. This study is being conducted by Mathematica Policy Research, Inc.

[6]Collection of data on race and Hispanic ethnicity by these questions will become a requirement in October 2004.

not clear to CMS how the state reports the data for the MSIS categories, which are separate for Pacific Islander and Asian status.

Since Medicaid eligibility is based in part on low-income status, only those below the income eligibility or "medically needy" thresholds in each state qualify for the program and are included in the data system.[7] Income and resource information is collected during the enrollment process and so MSIS includes a field for enrollees' income and for information on resources if states have further resource tests for eligibility. However, these data are not consistently reported to the federal government. MSIS does not contain a field to collect information on language use.

There are no requirements to report individual data for SCHIP, although states are required to report aggregate information to the federal government. If states use SCHIP funds to expand their Medicaid programs, then individual enrollment data (including racial and ethnic data) are reported to the federal government through MSIS. While states are required to report race and ethnicity in the aggregate, only about one third of all states do so (personal communication with CMS staff). Because there are no standard racial and ethnic categories for the enrollment forms, the quality of the data aggregated at the state level is suspect. Increased cooperation between the Medicaid and public health agencies would provide many states with opportunities to coordinate data collection goals and interventions to improve disparities in the health status of racial and ethnic minority groups. SCHIP does collect income information on its enrollment forms because it is needed to determine eligibility. However, eligibility rules differ from state to state, and there are no standardized ways to collect the data. These data are not reported to the federal government either. No language or acculturation data are collected for federal reporting of SCHIP data.

RECOMMENDATIONS

State and local governments maintain a wide array of data systems that can be used to better understand the health and health care of their constituent populations. Many of these data systems also play a crucial role in providing data for national-level health and health care research. Because these state-based data systems provide a substantial portion of the data

---

[7]These thresholds vary from state to state and across eligibility category (e.g., infants, children, pregnant women, and the elderly). For example, income thresholds for infants range from 133 percent of federal poverty levels (in many states) to 300 percent of federal poverty levels in New Hampshire in 2002 (National Governors Association, 2003). Income eligibility for children aged 6 to 18 ranged from 100 percent of federal poverty levels in many states to 275 percent of federal poverty levels in Minnesota in 2002.

available on health and health care services, they have great potential for furthering the understanding of disparities in health and health care and for the design and evaluation of interventions to eliminate them.

But the collection of data on race and ethnicity in these state-based systems is inconsistent. While the Medicaid program, through its new MSIS, will now collect data on race and ethnicity in a consistent way, SCHIP does not. Hospital discharge systems also do not uniformly collect racial and ethnic data across states; some states mandate the collection of these data as a part of their hospital discharge data reporting, others do not. Racial and ethnic data are not consistently collected in cancer registries, either. Both the SEER and NPCR systems rely on medical records for the information gathered in these systems. The racial and ethnic data that exist in these medical records are used to aggregate individuals into broad racial and ethnic classifications that are not consistent with the OMB standards. Thus, many opportunities to use these state-based systems for research on disparities in health and health care are missed because states do not require that the data be collected in these systems or that they be collected in a consistent manner.

State requirements to collect data on race and ethnicity in these systems could improve states' abilities not only to identify and monitor disparities in health and health care but also, more importantly, to design and evaluate intervention programs to eliminate the disparities. States can tailor their own requirements to ensure that data are collected on specific populations of interest in their states. By requiring standardized reporting of racial and ethnic data, states may also improve the completeness and consistency of the data collected. For example, not surprisingly, states that mandate the collection of data on race and ethnicity on hospital discharge abstracts have more complete reporting of these data than states without mandatory reporting (97 percent of abstracts in states with mandatory reporting contained data on race and ethnicity, compared with 83 percent in states with nonmandatory reporting) (AHRQ, 1999).[8]

At the same time, state requirements for the collection of racial and ethnic data would benefit research on health and health care disparities at the national level. Data from hospital discharge abstracts and from cancer registries are important especially for research on disparities in health care, an area where federal sources of data are lacking. Data from Medicaid and SCHIP, if they can be linked with claims data as the MSIS hopes to do, can

---

[8]While mandates may increase the completeness of these records, differences in the accuracy of data reported under mandatory versus nonmandatory systems have not been examined.

also provide vital information on the health care of low-income individuals as well as for different racial and ethnic groups.

The collection of data on SEP and language use and acculturation is not standardized in state-based data collection systems and these data are, in fact, rarely collected at all. This is probably in part because the settings in which data from vital records, hospital discharge abstracts, and cancer registries are collected are often not well suited for collecting extensive amounts of data, especially for concepts as difficult to measure as economic resources and degree of acculturation. However, some simple measures of SEP, such as education level or occupation, could be collected fairly easily in these data systems. In addition, states should examine the feasibility and potential usefulness of collecting simple measures of language use and acculturation, such as place of birth or generation status.

> RECOMMENDATION 5-1: States should require, at a minimum, the collection of data on race, ethnicity, socioeconomic position, and, where feasible, acculturation and language use.

The panel recognizes that there are barriers to states imposing data standards where none existed before. For example, there are costs to changing reporting and computer systems when new standards are imposed. Further, the actual recording of an individual's racial and ethnic data is often done by those who are not survey interviewers by training but rather work in health care or program administration (e.g., medical records clerks, providers and health care workers, or funeral directors). Some training in recording these data is often warranted, however, so that they are consistently captured. For instance, admitting clerks or eligibility workers may code a client's race and ethnicity through observation, making an assumption based on external characteristics such as surname, accent, skin color, or other features. Others may be trained to ask patients directly for their race and ethnicity, using a form or verbal interview. But training frontline workers is not a one-time effort—some of these positions experience high turnover rates so that the need for training is almost constant. Finally, getting political support for the collection of data on race and ethnicity during the health plan enrollment process or health care encounters requires education of both health system administrators and the general public about the benefits and risks of capturing and providing these data, and clear explanations of how they will be used and not used are essential. DHHS should therefore provide states with guidance and support to educate the public about the benefits of the data and to train administrative and medical records personnel to improve the recording of racial and ethnic data.

Many data collection systems have incomplete data because respondents refuse to answer questions about race, ethnicity, language, or SEP or because recorders fail to request or ascertain the information. How these missing data should be handled in analysis (e.g., by imputation or by supplementation with other data) is an issue on which many states need guidance. States should in any case continue to implement the new OMB standards for the collection of racial and ethnic data as these standards provide both a framework for collecting the data and the flexibility to capture more detailed data.

That said, several technical issues arise from the implementation of the standards. First, states need to serve their own populations first and foremost and so the categories of race and ethnicity that they use may or may not need to be more specific than the minimum OMB categories. However, there are great benefits to implementing at least the minimum categories for promoting multistate and national-level analyses of these data. Second, many of the data systems maintained by states derive from federal programs (Medicaid and SCHIP) or are part of federal/state cooperative data collection programs (e.g., vital records, NPCR, and SEER). Medicaid, SCHIP, and SEER do not require states to report data on race and ethnicity according to the OMB standards. Third, in order to understand changes in health disparities over time, comparisons across different reporting category systems will be necessary. The process of bridging the new categories to the old categories to allow comparisons of races and ethnicities over time is a technical issue that will need to be addressed. Furthermore, the new OMB standards allow individuals to choose multiple races and ethnicities in responding to questions, and how these multiple responses are converted into single category responses is another challenge that states will face. Much work on these technical issues has already been conducted by federal statistical agencies (Parker et al., 2003). DHHS should draw on this work to develop guidance for states on how to address these issues.

The collection of racial and ethnic data is beneficial only to the extent that the data are used to improve quality of care, administration of state programs, and the broad health status of state populations. It is important for key stakeholders in the data collection process to see that the data are being used for important and appropriate purposes. Individuals providing data are likely to feel better about doing so if they see that the data are being used in beneficial ways. Medical records clerks and providers that collect the data will also want to see them used so that they know their efforts are worthwhile. For all these reasons it is important, for both state and federal government collection of racial and ethnic data, that the usage of these data in reports and policy analysis be publicized. The value of the data also increases if they are made widely available to those who can use them. Health departments and policymakers are obviously the first layer of users at the state level, but researchers and analysts from a variety of settings (academia, federal government, local governments, and advocacy groups) could also use the data to effectively improve knowledge of health disparities. At the same time, privacy and confidentiality protections should be in place and enforced to guard against improper uses.

Some states have much experience in the collection, use, and dissemination of data on race, ethnicity, SEP, and language use and acculturation. Other states do not and could learn much from those that do. DHHS can exercise leadership in encouraging states' collection and use of existing data. DHHS uses the data it collects to produce a number of reports on health disparities, including such efforts as Healthy People 2010 and the National Healthcare Disparities Report. These and other efforts by DHHS agencies to implement research programs to better understand health disparities can serve as useful illustrations of the importance of how data on race, ethnicity, language, and SEP can be used. The federal government can also facilitate the sharing of information about how different states that collect the data use them in policy environments and make them available for outside users. Through these examples as well as the convening of conferences and the provision of training and technical assistance, DHHS could exemplify beneficial uses of this type of data.

RECOMMENDATION 5-2: DHHS should provide guidance and technical assistance to states for the collection and use of data on race, ethnicity, socioeconomic position, and acculturation and language use.

DHHS can consider a number of ways to support states in their collection of these data. One possible way for DHHS to encourage states to collect and use racial and ethnic data is by providing incentives to states in the form of competitive matching grants that target data collection for interventions to minority groups. An example of one such federal-state partnership is the Title V Maternal-Child Health Program (MCH), which provides matching funds to states and requires quantifiable outcomes monitoring, supported by the state data systems. One of the core MCH measures required by Title V is the Black and White Infant Mortality Ratio Performance Measure. The Vital Statistics Cooperative Program is another example of a mechanism with which DHHS can encourage states to collect and use racial and ethnic data in a standardized way. In this program, states are funded by the NCHS to provide standardized data on basic data items cooperatively agreed to by states and the NCHS.

To address gaps in coverage in many administration records, some states link public health data sets. For example, researchers from the New Hampshire State Department of Health and Human Services link data from the New Hampshire State Cancer Registry to state hospital discharge data (Taylor and Liu, 2003). In Ohio, Koroukian, Cooper, and Rimm (2003) linked Medicaid claims data with state cancer registry data to study the accuracy of Medicaid claims data in identifying breast cancer cases. Other states link Medicaid enrollment and claims data with hospital discharge data and cancer registries; Medicaid enrollment data with birth certificate data to understand prenatal and birth outcomes; immunization registry data with data from the Special Supplemental Nutritional Program for Women, Infants, and Children (WIC) and from birth records to fill in data gaps in the registry; and immunization registries with Medicaid and health insurance plan enrollment files for Health Plan Employer Data and Information Set measures. Although none of these linkages is done specifically to fill in gaps in information on race, ethnicity, SEP, or acculturation and language use, similar linkages could be attempted to use the strengths of some data sources in measuring these characteristics to supplement what is not collected in other data sources.

In order to meet the data collection and analysis needs for measuring and reducing disparities, the public and private sectors will need to coordinate efforts to develop state data and research agendas and design effective interventions. No sector can address this by itself.

6

Private-Sector Collection of Data on Race, Ethnicity, Socioeconomic Position, and Acculturation and Language Use

When an individual sees a physician or checks into a hospital, basic information on his or her health conditions and history, demographics, and insurance status is collected to provide physicians, nurses, and health care workers with health status and health care information and to process payment for services. Likewise, when an individual or family applies directly for health insurance or uses health insurance benefits, the health insurance company collects basic background information on the individual or family members and may use it to underwrite the insurance policy, to target outreach efforts that improve health and health care knowledge or aid in disease management, and to better understand the needs and health care services usage of the insurer's enrollees.

Data from these systems are collected as part of a health service interaction, but they are also used for statistical purposes—that is, to understand and draw inferences about how health care services are utilized, who is utilizing them, and the effect treatments have on health status, among many other questions. Such private-sector record systems can provide a rich set of data to understand disparities. These records are especially important sources of data for understanding health care disparities because they contain information on health care treatment (diagnoses, services received, procedure codes, and billed charges) that could not be collected in a survey setting without significant costs and respondent burden. Data collected by hospitals are aggregated at the state and federal level for these statistical purposes; data collected by physicians and by health plans are aggregated to a lesser extent.

In this chapter, we describe some of these private-sector data systems and their approach to the collection of data on race, ethnicity, socioeconomic position (SEP), and language. Private-sector data collections do not fall under the same regulatory framework as federal and state-based data collection, and in this chapter we briefly review the legal environment for private-sector collection of racial and ethnic data. We then review which data on race, ethnicity, SEP, and language are collected by health insurers, providers and medical groups, and hospitals, and barriers to the collection of these data. The chapter ends with the panel's conclusions and recommendations for the improvement of these data collection systems. In these recommendations, the panel encourages the Department of Health and Human Services to press for private-sector collection of data on race, ethnicity, socioeconomic position, and language use, to exercise leadership in setting standards for the collection of these data, and to develop mechanisms to make data linkages possible. Although these recommendations are made to promote data collection for statistical and research purposes in order to better understand and design programs to eliminate disparities in health and health care, the chapter also highlights how such data could be used by those organizations collecting the data to improve programs, services, outreach and treatment for individuals and groups to monitor and reduce disparities.

PRIVATE-SECTOR DATA COLLECTION SYSTEMS

We define private-sector data systems as those that collect data as part of a patient encounter with a medical professional at a hospital, clinic, nursing home, or medical group practice,[1] as well as those that collect data as part of private health insurance enrollment or a claims submission process (not including enrollment or claims data from Medicare, Medicaid, or State Children's Health Insurance Program [SCHIP]). In general, these data come from three sources: hospitals and nursing homes, health insurers,[2] and private physicians and physician medical groups. It should be noted that these private record collection systems sometimes share data on a limited basis with states and with federal data systems; for example, the National Hospital Discharge Survey (NHDS), which was discussed in Chapter 4, is a federally sponsored survey of records collected by hospitals on

---

[1] We do not include records of encounters at Veterans Health Administration or VA hospitals in this chapter, although data on veterans' use of health care services have been used to study racial and ethnic disparities in health care (for example, Jha et al., 2001). We also do not include records of patient encounters at Indian Health Service centers (see Chapter 4).

[2] Our definition of health insurers encompasses indemnity health insurers, managed care plans, health plans, and health maintenance organizations (HMOs).

inpatient care, and the two federal cancer registry systems—Surveillance, Epidemiology, and End Results system (SEER) and the National Program of Cancer Registries (NPCR)—collect some information from medical and laboratory records. Thus, the review in this chapter of the collection of data on race, ethnicity, SEP, and acculturation and language use by private record systems has implications for other data systems described in the previous chapters.

Hospitals collect information on patients during the admissions process. These data typically include background information on the patient (this information may include race, ethnicity, income, and/or education level); initial health conditions and symptoms, health insurance coverage, if any (including coverage through Medicaid, Medicare, or SCHIP); and services and treatment received while at the hospital. Information is kept on the patient's treatment during the hospital stay and compiled in a document that is completed when the patient is discharged, called the discharge abstract.[3] Many of these discharge abstracts are forwarded to state health agencies or to the federal government for statistical purposes. The National Hospital Discharge Survey collects medical records from a sample of nonfederal (which excludes military and VA hospitals), short-stay hospitals. The Healthcare Cost and Utilization Project collects hospital discharge data from 33 state organizations and generates a national data set, the Nationwide Inpatient Sample (NIS), which is a 20 percent sample of hospitals in the nation, and a State Inpatient Database (SID), which represents all acute-care discharges in the participating state.

There is some standardization of billing forms that hospitals use to bill for services provided. These include the Uniform Bill (UB92), which is currently used by hospitals, nursing facilities, and clinics to bill third-party insurers and government programs such as Medicaid, and the HCFA 1500, which is used to bill for professional services, such as physician or laboratory visits.

Health insurance data on individuals are collected when the individual both applies for health insurance (enrollment) and utilizes a health care service covered under the health insurance plan. Enrollment data include basic demographic information and may include medical history, employment status, and ability to pay premiums. Insurance claims forms typically include member (patient) identification information, dates of services, diagnoses, procedure codes, and billed charges.

Data collected from medical groups, physicians' offices, and group practices are similar to those collected by hospitals. They may include

---

[3]Information from birth certificates, disease reports, and cancer registries is also collected by hospitals; these were discussed in previous chapters.

general information on the patient's demographic characteristics (e.g., age, gender) as well as on treatments received, diagnoses, and payment status.

LEGAL FRAMEWORK

Private-sector data collection does not, in general, fall under the same rubric of laws and policies requiring the collection of racial and ethnic data as collections sponsored by the federal or state governments. Privately collected data that are part of federal programs (e.g., Medicare) do fall under the DHHS Inclusion Policy and the OMB standards thus apply. Furthermore, hospitals that seek reimbursement for Medicare or Medicaid fall under discrimination enforcement provisions of Title VI of the Civil Rights Act. Some states have laws regarding the collection of racial and ethnic data (see Youdelman, 2002, and NRC, 2003). The Health Insurance Portability and Accountability Act (HIPAA) and the Employee Retirement Income and Security Act (ERISA), both of which establish some national standards for the regulation of health care in the private sector, are other important parts of the legal backdrop for collecting racial and ethnic data that may affect private-sector health data collection. We briefly review the legal framework in this section.

The Civil Rights Act and Title VI

Title VI of the Civil Rights Act of 1964 prohibits discrimination on the basis of race or national origin in services rendered through federal programs. Compliance with Title VI is required in order for hospitals to receive reimbursement under Medicare and Medicaid programs,[4] although it does not apply to physicians' offices or group practices (see the paper by Nerenz and Currier in Appendix F).

The enforcement of the Civil Rights Act requires the documentation of the absence of discriminatory treatment. The law therefore requires hospitals (and health care providers, health plans, and other organizations that receive federal funds) to collect data and maintain records that can be used to monitor disparities (Perot and Youdelman, 2001) and show compliance with the law. Title VI does not, however, require any specific data collection. Rather, Department of Justice regulations that implemented Title VI require data collection to document compliance with the DOJ regulations[5] (DHHS regulations on Title VI are less explicit). The courts have ruled that

---

[4]See Smith (1999) for a useful review of regulations and enforcement regarding Title VI and Medicare.
[5]28 C.F.F. §§ 42.404, 406.

specification and enforcement of requirements for data collection under Title VI are at the discretion of federal agencies that run the programs covered by the act.[6]

Title VI also prohibits discrimination on the basis of national origin. Because a person's primary language has been accepted as a proxy for national origin, primary language data collection could be part of Title VI enforcement (Youdelman, 2002).

The legality of the collection of data on race, ethnicity, acculturation and language use, and SEP by health insurers has received much attention from federal agencies, advocacy groups, and researchers. Such data collection is seen by some as an important source of information both for measuring and understanding disparities in health care utilization and treatment and for identifying opportunities for targeted preventive programs. Yet fears of compromising patient privacy or of redlining by health insurers (the practice of charging different prices or denying coverage to individuals based on race, ethnicity, or national origin) leave room for a great deal of uncertainty in perceptions of data collection. Although disclosing confidential information and redlining are both illegal, such fears remain (and are not necessarily unfounded).

A recent review of the legalities of collecting data on race, ethnicity, and language concluded that there are no federal laws or regulations barring the collection of these data by health insurers (Perot and Youdelman, 2001). Only four states (California, Maryland, New Hampshire, and New Jersey) explicitly prohibit or partially prohibit the collection of racial, ethnic, SEP, and language data on insurance application forms in the individual or group insurance market (National Health Law Program, in press).[7] Five other states have regulations that could discourage the collection of data on race, ethnicity, and SEP: in Connecticut, Iowa, Minnesota, South Dakota, and Washington, when a health plan seeks state approval for its application forms, those that ask about race or ethnicity may be either disapproved or scrutinized closely.

---

[6]In *Madison Hughes v. Shalala* plaintiffs alleged that they were discriminated against by health care providers who were recipients of DHHS funds, in violation of Title VI, which prohibits recipients of federal funding from discriminating on the grounds of race, color, or national origin. The 6th District Court decided that it lacked subject matter jurisdiction. It held that there is a distinction between the department at its discretion requiring the collection of data to enforce Title VI and a requirement under the law that the department should use nationwide statistics to enforce Title VI. 80 F.3d 1121, 1124-1125 (6th Cir. 1996).

[7]We note that in October 2003 California voters rejected a ballot initiative that would have restricted the collection of data on race and ethnicity by the government except for medical research and certain law enforcement procedures.

Health Insurance Portability and Accountability Act

Another component of the legal backdrop for the collection of data on race, ethnicity, and language from private-sector sources is HIPAA. HIPAA, among other things, imposes national standards for electronic transactions with which all health insurers, health care clearinghouses, and providers (e.g., hospitals and physicians) that conduct business electronically must comply. The law requires DHHS to use the HIPAA transaction and code set standards for all covered transactions, including certain claims and enrollments. Transaction sets help standardize the business interactions among health care providers, health plan payers, and health plan sponsors; code sets define the data element values used in the standard transactions. Every time health care providers electronically transmit a claim to an insurer and in turn to the Centers for Medicare and Medicaid Services (CMS), HIPAA requires the use of adopted standard transactions.

Under HIPAA, Designated Standards Maintenance Organizations (DSMOs) are responsible for maintaining the content and standards for covered transactions according to Implementation Guidelines. These guides define the content and code sets of standardized data for constructing a HIPAA-compliant transaction (45 CFR Part 162, see U.S. DHHS, 2002), as well as the elements that providers and insurance plans must report in electronic health care claims and benefits enrollment transactions.

The HIPAA Implementation Guidelines state that each element in a transaction standard must be designated by the industry as either *required*, *situational*, or *not used* for completion of the transaction. Under this classification system, *required* elements must be included on the standard transaction, *situational* elements are used in certain circumstances but not others and *not used* elements are not currently reported. In the health care claims and benefits enrollment transaction standards, the fields for race and ethnicity are currently designated as *not used* and therefore are not among the HIPAA requirements for data collection.

The final adopted standards were published in the *Federal Register* on February 20, 2003.[8] Although data on race and ethnicity are not part of the HIPAA requirements for data collection under these adopted standards, there is a mechanism through which such data could become required. The DSMOs, in consultation with other committees with interests in the standards, called the Data Content Committees (DCCs),[9] can recommend

---

[8] See http://www.cms.hhs.gov/regulations/hipaa/cms0003-5/0003ofr2-10.pdf.
[9] These DCCs include the National Uniform Billing Committee, the National Uniform Claim Committee, the Workgroup for Electronic Data Interchange, and the American Dental Association.

changes to the standards. The secretary of DHHS can also modify any established standard, with some limitations regarding the criteria for which modifications can be made and the frequency of modifications (45 CFR Part 162, 2002). Any change the secretary proposes must be considered in consultation with the DCCs and, where relevant, the National Committee on Vital and Health Statistics.

CURRENT PRACTICES

This section describes current practices in collecting data on race, ethnicity, acculturation and language, and SEP by hospitals, health insurers, and medical groups, and identifies barriers to the collection of these data. The discussion is informed by two papers commissioned by the panel, both of which are included in the appendixes to this volume. Nerenz and Currier review racial and ethnic data collected by hospitals, health plans, and medical group practices. Bocchino discusses the results of interviews with health plans regarding their racial and ethnic data collection practices.

Data Collection by Hospitals

Hospitals play a major role in a community's health care delivery system and the health of its workforce. As communities become more diverse, hospitals are challenged to enhance their capacity to design and implement programs and treatment protocols that reduce or eliminate health disparities. Racial and ethnic data provide an important foundation for designing such programs and protocols. There is some evidence that hospitals have recognized the need for the collection of data on race and ethnicity. The Health Research and Education Trust (HRET—an affiliate of the American Hospital Association), in collaboration with Michigan State University, surveyed a nationally representative sample of hospitals regarding their practices in collecting data on race and ethnicity (see the paper by Nerenz and Currier in Appendix F). Results from this survey show that hospitals collect these data for a number of reasons—for example:

- to use for quality improvement measures,
- to target hospital marketing efforts,
- to fulfill requirements by law or regulation, and
- to improve community relations.

The HRET and Michigan State survey provided a snapshot of data collection practices showing that 79 percent of responding hospitals said they collected racial and ethnic data on their patients.[10] The patient was the source of this information in a majority of hospitals, but a significant percent of hospitals reported that clerks code patient race and ethnicity by observation.

The infrastructure for collecting and using data on race and ethnicity in the hospital industry is undeveloped and faces several barriers. Because racial, ethnic, and other socioeconomic data are not needed to pay a claim, they are not universally or uniformly collected in hospital discharge abstracts, which are the major source of hospital patient diagnostic and treatment data for research and public health statistics. This lack of a standardized approach to the collection of patient's racial, ethnic, and socioeconomic data is a major barrier to understanding and eliminating disparities, not only for hospitals but also for state and national data collection systems that rely on the data collected by local hospitals. The result is the underreporting of information on race and ethnicity, variability in methods of capturing the data, and misclassification. A New York study compared records of individuals with two separate admissions to hospitals and found that they agreed 93 percent of the time, but that agreement for groups other than blacks and whites were lower (Blustein, 1994).

The lack of uniformity in data collection among hospitals that do capture racial and ethnic data is a barrier to making comparisons across providers and communities. Results from the HRET survey (discussed in the paper by Nerenz and Currier in Appendix F) show that while the majority of hospitals are collecting data on race, ethnicity, and language, they are collecting it very differently—findings that are consistent with other studies that document the quality of racial and ethnic data collection by hospitals. The HRET survey discovered that the majority (70 percent) of responding hospitals that collect racial and ethnic data did not see any drawbacks to collecting the data. However, drawbacks of data collection cited by the remaining 30 percent include:

- problems associated with the quality and accuracy of the data,
- discomfort of admitting clerk in asking the patient for the information,
- concerns that patients might be insulted or offended if asked about their race and ethnicity,
- patients often did not fit the given racial or ethnic categories,
- fears that the data might not be kept confidential, and

---

[10]A total of 262 of 1,000 hospitals responded. It is difficult to assess the generalizability of these results given the low response rate.

• concerns that the collection of racial and ethnic data might be used to profile patients and discriminate in the provision of care.

Of hospitals not collecting racial and ethnic data, the majority cited the belief that it was unnecessary. Other reasons given for not collecting the information included the time and resources involved in collecting and managing the data and concerns about the classification system.

A convergence of national awareness about health disparities and the increasing market and policy incentives for the hospital industry to respond have resulted in two important industry initiatives:

• HRET has been working with a consortium of six hospitals and health systems to develop a uniform framework for collecting racial, ethnic, and primary language data in hospitals. As part of this effort, HRET has conducted site visits and, in collaboration with Michigan State University, conducted a survey of hospital data collection and use practices. The goal of this initiative is to inform the development of a systematic and uniform framework for the collection and use of data on race, ethnicity, and language data across hospitals.

• The American Hospital Association (AHA) has added two new questions to its annual survey, which is a survey of the more than 6,000 AHA member hospitals. Beginning in 2003, hospitals were asked whether they collect information on the patient's race, ethnicity, and primary language spoken. This information will be used to provide important baseline and trend information about data collection practices across hospitals over time.

The first of these two initiatives will be particularly useful for illuminating how the barriers to collecting these data (such as those identified by hospitals in the paper by Nerenz and Currier in Appendix F) are experienced and ultimately addressed.

Data Collection by Health Insurers

Very few health insurance companies collect data on race, ethnicity, acculturation and language, and SEP. A few recent studies have surveyed health insurers to learn about their practices in collecting these data. For example, Bocchino in Appendix G, interviewed 30 health insurers who were members of the American Association of Health Plans about their racial and ethnic data collection practices.[11] This study found that of these

---

[11]See her paper in Appendix G. Sixteen of these insurers were chosen for interview because they were known to have initiated race- and ethnicity-related projects. The remaining 14 were randomly selected from the 2002 AAHP Industry Survey respondents lists.

30 insurers, a quarter of them asked about race and ethnicity on enrollment forms completed by accepted applicants, but that these voluntary questions were frequently left unanswered.[12] It was unclear from the study whether the data were collected solely for those requesting individual or family coverage or for employer-based coverage as well. In another small survey, Nerenz and Currier found that while few insurers collect racial and ethnic data, those that do use the information to prepare language translations of materials, for quality improvement purposes, and to inform disease management programs. Nerenz and Currier also report, however, that an informal survey in 1999 of large not-for-profit health insurers involved in the Health Maintenance Organization Research Network in 1999 found that most insurers did not collect data on the race and ethnicity of their members.

Bocchino reports that health insurers' enrollment forms have included questions on language preference for a number of years, but that these fields are usually optional and are often left blank. One major health insurer shared with the panel its data collection practices for various insurance products. For some (but not all) products, this insurer collects primary language usage data from its members on an optional basis on application forms. For two products, this insurer collected salary information on application forms, but this was an exception to its standard practice. Otherwise, no other SEP data were collected by the groups in this plan.

Awareness of the potential utility of the collection of racial and ethnic data by health insurers was raised recently when Aetna, which insures 14 million individuals, announced that it would begin collecting racial and ethnic data from its members either once they were accepted for coverage or when they requested a change in coverage (Winslow, 2003). Since the initiative began in September 2002, about 64,000 enrollees in 13 states and the District of Columbia have been asked to voluntarily indicate their race, ethnicity, and language preference, and about 80 percent have provided the information.[13] Aetna reports that the data will be used to "create more culturally focused disease management and wellness programs for our multicultural membership."[14] Aetna is the first health plan to publicly announce such an effort.

Other health insurers have initiated more limited efforts on racial and ethnic data collection and in support of cultural competence activities. These efforts and activities include the formation of CEO-level task forces,

---

[12]Enrollment in a plan and application to a plan are different. Enrollment occurs after an insurer has accepted the application for coverage.

[13]Collection of this information in four more states was recently added to this effort.

[14]Quality Care for All: Reducing Racial and Ethnic Disparities in Health Care: A Special Interest Publication for America's Employers.

the collection of racial and ethnic data on membership satisfaction surveys, the use of language preference data from enrollment forms to target health plan materials, and collaboration with state public health departments to link member files with state public health data files and surveys (as discussed in the paper by Bocchino).

Many health insurers report interest in collecting these data to target prevention and treatment programs. In her survey of AAHP health insurers, Bocchino reports that data on race, ethnicity, language preference, and other individual characteristics such as sex, age, education, and geographic location are important for health insurers to appropriately target information and programs to improve the health of enrolled members. In Bocchino's study, health insurers that collect these data reported that when the data were provided by enrollees and recorded on their enrollment forms, they were in fact useful for identifying populations at higher risk for chronic conditions and for targeting appropriate preventive care programs. Health insurers have also used these data to support disease management activities.

There are, however, barriers to the collection of racial, ethnic, acculturation and language, and SEP data by health insurers. Perhaps the most significant obstacles are concerns about the legalities of collecting these data and about perceptions among those who provide the information that the data might be used for discriminatory purposes. In the Bocchino study, almost two-thirds of the responding insurers cited legal concerns as the most important deterrent to collection. Fremont and Lurie (see their paper in Appendix D) indicate that some plans fear that collecting such data would increase their exposure to litigation over potential privacy violations or violations of civil rights laws, or that potential members, or the employers who choose health insurers for their employees, would respond negatively to the collection of these data for fear that they may be used detrimentally.

Given concerns about perceived reasons for collecting racial and ethnic data at enrollment, several alternative methods of data collection have been suggested. One alternative is to collect these data on members once they have already enrolled, an approach that may allay fears that the data will be used to discriminate in the underwriting process. Another alternative is for HMOs and preferred provider organizations (products that have defined physician and hospital networks) to ask providers to collect these data at the point of service. However, some of the standardization and perception problems discussed above with regard to collecting these data on hospital medical records apply for this type of data collection also, since providers may be hesitant to ask these questions of patients, fearing the same perception problems and the perception of increased litigation exposure. Collecting such data at the point of service would mean that the data would need to be collected repeatedly—every time an individual used a service—whereas it would be less burdensome for insurers to collect the data only once. Furthermore, collection at the point of service would mean that data would not be available for the large number of health members who are insured but have not used their coverage (Fiscella, 2002) and that information on race and ethnicity would be collected only on the subset of members who submit claims. Any characterization of the health insurers' membership would thus be limited to those seeking care, or roughly two-thirds of a typical insurer's membership. This would limit the generalizability of study results. A third option may be to conduct a separate survey of already enrolled members to collect these data. High costs and quality of the data are important considerations for this type of collection.

Health insurers interviewed by Bocchino cited the costs of collecting information and of updating systems to collect and store the information as a barrier, as well as the requirement for approval by various government agencies of any revisions on enrollment forms (e.g., a state insurance department).

Data Collection by Medical Groups

Very little is known about the collection of racial, ethnic, language, and SEP data by private physicians' offices and group practices. While Nerenz and Currier found that some practices collect racial and ethnic data on their patients, the information is usually collected for their own purposes (e.g., internal quality improvement and disease management activities) and is therefore not standardized, consistently collected, or made available for public use.

Quality of Care Measures

The Consumer Assessment of Health Plans Survey (CAHPS) and the Health Plan Employer Data and Information Set (HEDIS) quality of care measures are systems of quality measurement that are widely used by health plans for their HMO products.[15] Both of them are required as part of the voluntary accreditation process administered by the National Committee for Quality Assurance (NCQA), which sponsors development of the HEDIS measures. Many purchasers—including Medicare, the Federal Employees Health Benefits Program (FEHBP), and many private purchasers and state Medicaid programs—require reporting of one or both of these quality measures.

[15]The term *health plans* refers to insurers who offer HMO and preferred provider organization (PPO) products.

CAHPS was developed under a program sponsored by the Agency for Healthcare Research and Quality (AHRQ) with the objective of producing a standardized instrument for surveys of health plan members.[16] CAHPS consists of a core of items that are widely applicable, together with several sets of supplementary items designed for special populations (e.g., children, Medicare beneficiaries, people with chronic diseases). The CAHPS quality items ask for overall ratings of care as well as for reports on more specific aspects of respondents' experiences, such as waiting times or ease of obtaining particular services. The survey typically also collects very limited information on health status or conditions. Although it is primarily designed for HMO members, versions of the survey have been used to evaluate care in the Medicare fee-for-service sector and are being developed to evaluate medical groups, hospitals, and nursing homes. Since its introduction in 1998, CAHPS has been widely implemented, and therein lies its primary importance to this discussion, although much of the following description of its content is also applicable to the various other surveys used in hospitals, medical groups, and other health care institutions.

CAHPS includes limited content on race, ethnicity, SEP, and language use. In the standard instrument, race and ethnicity are measured by items that closely follow the standard OMB categories. The survey has been translated into Spanish and some other languages, and when these translated versions are used, they can serve as an indication of the respondent's language preference. Concerns have been raised, however, about the precise equivalence of Spanish- and English-language versions of the survey.

The only SEP measure is an item on education. Education is widely used for "case mix" adjustment—that is, to adjust for the component of scores that is attributable to differences in the composition of the membership of the different insurers rather than to differences in quality. More-educated (presumably higher-SEP) members tend to give lower ratings, likely reflecting higher expectations rather than poorer care. The racial and ethnic measures, on the other hand, are not typically used for adjustment, but have been used for subgroup analyses. Geographical linkages to CAHPS data are also possible. Although the main purpose of CAHPS is to support comparative reporting on health insurers, individual-level data are collected by some survey sponsors and are compiled on a voluntary basis in the National CAHPS Benchmarking Database to be used in research.

The HEDIS quality of care measures are used to estimate rates of provision of selected screening and preventive and chronic disease treatment services to eligible populations, based on a combination of administrative data and medical record reviews. Thus they collect no data on race,

---

[16]See http://www.ahcpr.gov/qual/cahpsix.htm for a description of this survey.

ethnicity, or SEP beyond those already included in these records. Analyses of race, ethnicity, and SEP effects in HEDIS measures have been conducted either using Medicare administrative data on race (Schneider, Zaslavsky, and Epstein, 2002) or through geographic linkages (Zaslavsky et al., 2000).

RECOMMENDATIONS

Data collected by private-sector groups could be invaluable sources for better understanding disparities in health care. These data could be used for statistical purposes by governments and private researchers to monitor the status and understand the causes of disparities in health care. As discussed by Fremont and Lurie, the data may also be directly useful to insurers, medical groups, and hospitals, which could use them to monitor differences in utilization of health care services.

For example, differences in utilization among members of the same health plan with the same coverage may reveal areas where improved quality of care is needed. Many health insurers provide targeted case management and support services for people with chronic conditions such as asthma, diabetes, and congestive heart failure. Racial and ethnic data could be used to develop culturally appropriate outreach for patient enrollment in these programs and ensure that follow-up support services are culturally appropriate. Moreover, data on race and ethnicity would enhance health insurer data sets and contribute to understanding disparities in preventive and palliative care among commercially insured populations. Finally, these enhanced data sets could be used to assess potential changes in service patterns associated with race or ethnicity among health care providers.

Hospitals could use data for quality improvement measures. Monitoring quality measures by race and ethnicity could identify areas for which quality improvement could be targeted. In addition, hospitals and large medical groups could use the data to gauge the need for services targeted to specific ethnic or minority groups (e.g., translators, educational materials) to improve quality of care. And consumers of health care services and health insurance could use information generated from these data, if it were publicly available, to make informed choices about the performance of services and health plans.

Educational programs developed by hospitals, insurers, or public health organizations may be more appropriately targeted to individuals and groups if information on race, ethnicity, and acculturation and language use are available to guide these efforts. Fremont and Lurie cite a study that found that a mass media campaign to educate the public about steps to take to avoid sudden infant death syndrome was less effective among black mothers than among white mothers because the messages were not appropriately targeted to black women (Malloy, 1998).

The panel's review of current practices by the private sector—hospitals, health insurers, and medical group practices—has revealed that the collection of data on race, ethnicity, language, and SEP is not common and not standardized. When hospitals collect racial and ethnic data on their patients, reports show that the reporting is fairly complete. However, the data are not reported in a standardized format and accuracy for groups other than white and black is suspect. Few health insurers collect data on race, ethnicity, and language. For those that do, individuals often do not provide their information. Finally, even less is known about the racial and ethnic data collected by medical groups. The collection of SEP data is probably even more rare in these privately based data collection systems; the only such data collected by hospitals is the source of payment for patients. Health insurers rarely collect information on education or income.

The panel believes that an opportunity to learn more about disparities is missed because private medical and insurance organizations do not routinely collect information on race and ethnicity, acculturation and language, and SEP. The lack of data from these sources is a serious weakness in the current systems of health data collection. DHHS could remedy this problem by intervening to ensure that these data are collected uniformly. Health insurers and hospitals have expressed interest in collecting these data but worry that, without a federal or state mandate, the collection of the data will be greeted with suspicion. Federal leadership is needed to help legitimize and regularize the collection of these data across states and health systems.

RECOMMENDATION 6-1: DHHS should require health insurers, hospitals, and private medical groups to collect data on race, ethnicity, socioeconomic position, and acculturation and language.

The panel does not have the expertise to assess whether DHHS has the statutory authority necessary to require private entities to collect data on race, ethnicity, acculturation and language use, and SEP. However, there appear to be several possible options for DHHS to pursue such requirements through existing laws, regulations, and initiatives.

HIPAA is one such vehicle, although it does not currently require the collection of these data. The race and ethnicity elements in the standard set of claims and enrollment transactions are currently designated as "not used" and thus are not reported. The secretary of DHHS is in a position to propose changes to the current HIPAA standards. Strong leadership from DHHS would be needed to guide the proposed changes through the process of approval from the DSMOs in consultation with other industry committees. The case for the proposed changes would be strengthened by the argument that the collection of racial and ethnic data is essential to meeting the Healthy People 2010 initiative to eliminate disparities in health and health care. Because HIPAA only covers electronic transactions, the addition of racial and ethnic data to the standards would not cover all transactions, but it would significantly enhance both the amount of data available for studying disparities and the effectiveness of interventions designed to eliminate disparities.[17]

A potential vehicle for standardizing the collection of data on race, ethnicity, SEP, and language use among hospitals is the Joint Commission on Accreditation of Healthcare Organizations (JCAHO), which is an independent nonprofit organization that is a standard-setting and accrediting body for more than 16,000 health care organizations. JCAHO already requires the collection of primary language information and could add requirements for the collection of racial and ethnic data.

Aligning incentives for hospitals to collect and use data on race and ethnicity is key to overcoming the barriers that now contribute to incomplete and non-comparable data. Given the importance of these data for public health and research, efforts at the hospital, governmental, regulatory, and national levels are essential for overcoming the barriers. The HRET and AHA efforts are to be commended and are a step in the right direction for improving the completeness and utility of racial, ethnic, and socioeconomic data collected by the nation's hospitals.

Creative ideas for overcoming concerns and disincentives to collect data on race, ethnicity, SEP, and acculturation and language use are needed. There are, however, examples of barriers that have been overcome. Leading hospital systems have proven that the collection of these data is possible. These hospitals have served as laboratories for collecting and using racial and ethnic data in a hospital environment. Despite the difficulties and limitations of the data, hospitals that made the investment have demonstrated the utility of the data by understanding—and improving services for—the community they serve. These hospitals can target quality improvement interventions and measure their effectiveness, comply with grant reporting requirements, and compete more effectively for research and service grants. They can also design their workforce to match the communities they serve and thus underscore their commitment to their mission and to their donors and communities.

Since the sources of these data are usually records rather than surveys, extensive data on socioeconomic position and acculturation and language use may be infeasible to collect without sizable costs and time commit-

---

[17]Although the Federal Employees Health Benefit Plan (FEHBP) is outside DHHS authority and is, rather, under the authority of the Office of Personnel Management of the U.S. government, another potential step forward is to require that FEHBP, which insures over 8 million federal employees and their dependents, collect these data.

ments. However, measures of education and occupation are more easily collected, as are proxy measures of acculturation and language such as place of birth, generation status, and primary language.

DHHS is a powerful player in health care transactions conducted by private entities through the Medicare and Medicaid programs. Racial and ethnic data for some Medicare enrollees are already available through Social Security Administration records. But as we described in Chapter 4, such data are not available for all enrollees, will not be available for some future enrollees, and are not always reported consistent with recent OMB standards for the collection of these data. The department could, through its administration of the Medicare program, require the collection of data on race, ethnicity, SEP, and primary language to fill in gaps. Providers, hospitals, and other entities seeking reimbursement for services provided under Medicare would then be required to provide data on the race and ethnicity of each individual who receives service.

DHHS could also promote standardized state-level collection of these data through each state's Medicaid program. Data on race and ethnicity are reported on the Medicaid enrollment forms, and the OMB standard categories are supposed to be used. However, DHHS could more strongly enforce the collection of these data and also offer guidance and technical assistance to help the states implement procedures to collect the data.

In collecting such data, hospitals, health plans, and medical groups should be aware that some individuals may be reluctant to provide the information. Respondents should be informed that they are volunteering to provide these data and should also be informed about how the data will be used. This approach may help assuage fears about confidentiality breaches and may encourage individuals to provide the data.

Promoting Standardized Collection of Data

DHHS should work with hospitals and health insurers to determine the best way to collect standardized data, using the OMB standards for collecting racial and ethnic data as a base. Further detail may be required for some hospitals or health insurers that serve a large number of individuals from smaller population subgroups beyond the OMB standard categories. DHHS should also work with hospitals, health plans, and related groups to determine which SEP measures could reasonably be collected on enrollment or admissions forms. Collection of these data will necessarily be limited as extended collection of wealth and income data is not feasible for these record systems. Education level may be the most practical item to collect and the least sensitive for individuals to provide.

In setting up data systems and standards for the collection of such data, DHHS and industry agents should try to design systems that avoid repeatedly collecting the same information on individuals. This will reduce the burden both for respondents and for those collecting the information.

In developing standards for data collection, it is also critically important to provide clear information that indicates how the data will be used and that the data are provided on a voluntary basis. Providing this information can help alleviate fears that the data will be used for discriminatory purposes. This information should be provided at the data collection point, which in most cases would be when the patients and plan enrollees fill out forms. Acknowledging the risks associated with the collection and use of data on race and ethnicity is part of the due diligence of the collection of these data by hospitals, health plans, and medical groups. Building trust by protecting the data from improper use or disclosure is essential. If the patients are told that providing their socioeconomic and demographic data will result in more translation services or community prevention programs, then these should be implemented. DHHS should work with industry agents and legal experts to develop the information to be given to individuals who are asked to provide the data.

> **RECOMMENDATION 6-2: DHHS should provide leadership in developing standards for collecting data on race, ethnicity, socioeconomic position, and acculturation and language use by health insurers, hospitals, and private medical groups.**

Linking Geocoded Data from the Private Sector to Federal Data

As noted throughout this chapter, only very limited data on race and ethnicity are typically collected in private-sector health care information systems. Implementation of this report's recommendations would greatly enhance the data infrastructure available for understanding and eliminating disparities. However, if these recommendations cannot be implemented such that high-quality data are produced, linking aggregate-level data on race, ethnicity, SEP, and language use may be needed to bridge the gaps. In general, provider, hospital, and insurance claim forms contain the claimant's address. The Bureau of the Census provides aggregated data on race, ethnicity, and SEP for census geographical units (Zip Code tabulation areas, tracts, or block groups), and these aggregated data, when geographically linked with data from private-sector records, can be used as proxies for individual data on race, ethnicity, or SEP.

Two technical issues are critical to implementation of such linkages. First, software must be used to link addresses to census geographical units. Second, census data must be linked to the addresses, with suitable protections for confidentiality. If the data will be disseminated beyond the plan for research purposes, the second step requires special care because the precise combination of values of the sociodemographic variables might identify the subject's geographical area and thus pose a risk of disclosure of confidential information about individual plan members. Methods have been developed for masking such data by rounding and/or adding random noise. Such masked data sets can be analyzed with appropriate corrections for the effects of masking. But development of the specific procedures and parameters required to implement data masking requires particular statistical expertise that is not likely to be found within health insurers. Considerable resources would be required to accomplish it. Furthermore, a uniform procedure should be followed so that data will be comparable across the private-sector units generating the data.

DHHS could greatly facilitate the routine generation of high-quality, uniform, and nondisclosing geographically linked data sets by providing a linking service that could be used by private- and public-sector health care organizations. Such a service could be administered, for example, through a Web site. An organization would anonymously submit a file containing member addresses and would receive in return a file of masked geographical variables at several levels. Although geocoding is an imperfect process, typically 85 percent of addresses in a health care file might be geocoded down to the block group level; cases that cannot be geocoded might be either imputed or analyzed using variables aggregated to higher levels of geography.

The greatest expertise in the federal government for solving the problems involved in establishing such a service resides in the Bureau of the Census. Within DHHS, the NCHS has been a leader in dealing with confidentiality issues. Alternatively, a private-sector vendor with the necessary geocoding expertise could be recruited, although such vendors do not typically deal with the related confidentiality issues.

RECOMMENDATION 6-3: DHHS should establish a service that would geocode and link addresses of patients or health plan members to census data, with suitable protections of privacy, and make this service available to facilitate development of geographically linked analytic data sets.

References

Acheson, D.
　1998　*Independent Inquiry into Inequalities in Health*. London, England: Her Majesty's Stationery Office.
Agency for Healthcare Research and Quality
　1999　HCUP *1998 Data Availability Inventory*. Salt Lake City, UT: National Association of Health Data Organizations and Medstat.
Aguirre-Molina, M., C. Molina, and R.E. Zambrana, eds.
　2001　*Health Issues in the Latino Community*. San Francisco, CA: Jossey-Bass.
Aligne, C.A., P. Auinger, R.S. Byrd, and M. Weitzman
　2000　Risk factors for pediatric asthma: Contributions of poverty, race and urban residences. *American Journal of Respiratory and Critical Care Medicine* 162: 873-877.
American Association of Physical Anthropology
　1996　AAPA statement on biological aspects of race. *American Journal of Physical Anthropology* 101:569-570.
American Sociological Association
　2003　*The Importance of Collecting Data and Doing Social Scientific Research on Race*. Washington, DC: American Sociological Association
Anderson, J., M. Moeschberger, M.S. Chen, Jr., P. Kunn, M.E. Wewers, and R. Guthrie
　1993　An acculturation scale for Southeast Asians. *Social Psychiatry and Psychiatric Epidemiology* 28(3):134-141.
Arday, S.L., D.R. Arday, S. Monroe, and J. Zhang
　2000　HCFA's racial and ethnic data: Current accuracy and recent improvements. *Health Care Financing and Review* 21(4):107-116.
Balcazar, H., and Z. Qian
　2000　Immigrant families and sources of stress. Pp. 359-377 in *Families and Change: Coping with Stressful Life Events and Transitions* (Second Edition), Patrick C. McKenry and Sharon J. Price, eds. Thousand Oaks, CA: Sage Publications.

Balcazar, H., G. Peterson, and J. Cobas
  1996   Acculturation and health related risk behaviors among Mexican American preg-
          nant youth. *American Journal of Health Behavior* 20:425-433.
Bamshad, M.J., and S.E. Olson
  2003   Does race exist? *Scientific American* 289(6):78-85.
Barsky, R., J. Bound, K. Charles, and J. Lupton
  2002   Accounting for the black-white wealth gap: A nonparametric approach. *Journal of
          the American Statistical Association* 97:663-673.
Baumeister, L., K. Marchi, M. Pearl, R. Williams, and P. Braveman
  2000   The validity of information on "race" and "Hispanic ethnicity" in California birth
          certificate data. *Health Services Research* 35(4):869-883.
Becerra, J.E., C.J. Hogue, H.K. Atrash, and N. Perez
  1991   Infant mortality among Hispanics: A portrait of heterogeneity. *Journal of the
          American Medical Association* 265(2):217-221.
Berry, J.
  2003   Conceptual approaches to acculturation. Pp. 17-37 in *Acculturation: Advances in
          Theory, Measurement, and Applied Research*, K. Chun, P. Balls, and G. Marin,
          eds. Washington, DC: American Psychological Association.
Blustein, J.
  1994   The reliability of racial classifications in hospital discharge abstract data. *American
          Journal of Public Health* 84:1018-1021.
Bound, J., and A.B. Krueger
  1991   The extent of measurement error in longitudinal earnings data: Do two wrongs
          make a right? *Journal of Labor Economics* 9(1):1-24.
Burchard, E.G., E. Ziv, N. Coyle, S.L. Gomez, H. Tang, A. Karter, J.L. Mountain, E.J. Perez-
Stable, D. Sheppard, and N. Risch
  2003   The importance of race and ethnic background in biomedical research and clinical
          practice. *New England Journal of Medicine* 348(12):1170-1175.
Bureau of Labor Statistics
  1995   *A Test of Methods for Collecting Racial and Ethnic Information.* Washington, DC:
          Bureau of Labor Statistics.
Byrd, T., H. Balcazar, and R. Hummer
  2001   Acculturation and breastfeeding intention and practice in Hispanic women on the
          U.S.-Mexico border. *Ethnicity and Disease* 11:72-79.
Carter-Pokras, O., and C. Baquet
  2002   What is a health disparity? *Public Health Reports* 117:426-432.
Case, A., D. Lubotsky, and C. Paxson
  2002   Economic status and health in childhood: The origins of the gradient. *American
          Economic Review* 92(5):1308-1334.
Centers for Disease Control and Prevention
  2003   *National Diabetes Fact Sheet.* Available: http://www.cdc.gov/diabetes/pubs/
          estimates.htm#prev (September 16, 2003).
Centers for Medicare and Medicaid Services
  2003   *Health Care Indicators.* Available: http://www.cms.hhs.gov/statistics/health-indicators/
          default.asp (June 2004).
Cho, Y., and R.A. Hummer
  2000   Disability status differentials across fifteen Asian and Pacific Islander groups and
          the effect of nativity and duration of residence in the U.S. *Social Biology* 48(3-
          4):171-195.

Chun, K., P. Balls, and G. Marin, eds.
    2003   Acculturation: Advances in Theory, Measurement, and Applied Research. Washington, DC: American Psychological Association.
Cilke, J.
    1998   A Profile of Non-Filers. (OTA Paper #78, Office of Tax Analysis). Washington, DC: U.S. Department of Treasury.
Clark, L., and L. Hofsess
    1998   Acculturation. Pp. 37-59 in Handbook of Immigrant Health. New York: Plenum Press.
Cobas, J.A., H. Balcazar, M.B. Benin, V.M. Keith, and Y. Chong
    1996   Acculturation and low-birthweight infants among Latino women: A reanalysis of HHANES data with structural equation models. American Journal of Public Health 86(3):303-305.
Coder, J.
    1992   Using administrative record information to evaluate the quality of the income data collected by the SIPP. Pp. 265-306 in Proceedings of Statistics Canada Symposium 92: Design and Analysis of Longitudinal Surveys. Ottawa: Statistics Canada.
Cooper, R.S., and R. David
    1986   The biological concept of race and its application to public health and epidemiology. Journal of Health and Political Policy Law 11:97-116.
Cooper, R.S., J.S. Kaufman, and R. Ward
    2003   Race and genomics. New England Journal of Medicine 348(12):1166-1170.
Covinsky, K.E., L. Goldman, E.F. Cook, R. Oye, N. Desbiens, D. Reding, W. Fulkerson, A.F. Connors Jr., J. Lynn, and R.S. Phillips
    1994   The impact of serious illness on patients' families. Journal of the American Medical Association 272(23):1839-1844.
Crump, C., S. Lipsky, and B.A. Mueller
    1999   Adverse birth outcomes among Mexican-Americans: Are U.S.-born women at greater risk than Mexico-born women? Ethnicity and Health 4(1-2):29-34.
Cuellar, I., B. Arnold, and R. Maldonado
    1995   Acculturation rating scale for Mexican Americans-II: A revision of the original ARSMA scale. Hispanic Journal of Behavioral Sciences 17(3):275-304.
Currie, J., and M. Stabile
    2002   Socioeconomic Status and Health: Why is the Relationship Stronger for Older Children? (NBER Working Papers 9098). Cambridge, MA: National Bureau of Economic Research.
Deaton, A.
    2002   Policy implications of the gradient of health and wealth. Health Affairs 21(2):13-30.
Deaton, A., and C. Paxson
    2001   Mortality, Income, and Income Inequality Over Time in Britain and the United States. (NBER Working Papers 8534). Cambridge, MA: National Bureau of Economic Research.
Deyo, R.A., A.K. Diehl, H. Hazuda, and M.P. Stearn
    1985   A simple language-based acculturation scale for Mexican Americans: Validation and application to health care research. American Journal of Public Health 75:51-55.
Diamant, A.L., S.H. Babey, E.R. Brown, and N. Chawla
    2003   Diabetes in California: Findings from the 2001 California Health Interview Survey. UCLA Center for Health Policy Research and the California Endowment. Available: http://www.healthpolicy.ucla.edu/pubs/files/UCLA_Diabetes_Rpt_Final_R2.pdf.

Duncan, G.J., M.C. Daly, P. McDonough, and D.R. Williams
  2002 Optimal indicators of socioeconomic status for health research. *American Journal of Public Health* 92(7):1151-1157.
Dynan, K., J.S. Skinner, and S. Zeldes
  2004 Do the rich save more? *Journal of Political Economy* 112(2):397-444.
Elixhauser, A., R.M. Weinick, J.R. Betancourt, and R.M. Andrews
  2002 Differences between Hispanics and non-Hispanic whites in use of hospital procedures for cerebrovascular disease. *Ethnicity and Disease* 12(1):29-37.
English, P., M. Kharrazi, and S. Guendelman
  1997 Pregnancy outcome and risk factors in Mexican Americans: The effect of language use and mother's birthplace. *Ethnicity and Disease* 7:229-240.
Epstein, A.M., J.Z. Ayanian, J.H. Keogh, S.J. Noonan, N. Armistead, P.D. Cleary, J.S. Weissman, J.A. David-Kasdan, D. Carlson, J. Fuller, D. Marsh
  2000 Racial disparities in access to renal transplantation: Are they due to clinical appropriateness, underuse, or overuse? *New England Journal of Medicine* 343:1537-1544.
Escalante, A., J. Barrett, I del Rincon, J.E. Cornell, C.B. Phillips, and J.N. Katz
  2002 Disparity in total hip replacement affecting Hispanic Medicare beneficiaries. *Medical Care* 40(6):451-460.
Escarce, Jose J., and T. McGuire
  1993 Racial differences in the elderly's use of medical procedures and diagnostic tests. *American Journal of Public Health* 83(7):948-954.
Federal Interagency Committee on Measures of Educational Attainment
  2000 *Federal Measures of Educational Attainment: Report and Recommendations.* Washington, DC: Federal Interagency Committee on Measures of Educational Attainment.
Fennell, M.L, C. Miller, and V. Mor
  2000 Facility effects on racial differences in nursing home quality of care. *American Journal of Medical Quality* 15(4):174-181.
Fiscella, K.
  2002 Using existing measures to monitor minority health care quality. In *Improving Healthcare Quality for Minority Patients: Workshop Proceedings.* Washington, DC: The National Quality Forum.
Fiscella, K., P. Franks, M.P. Doescher, and B.G. Saver
  2002 Disparities in health care by race, ethnicity, and language among the insured: Findings from a national sample. *Medical Care* 40(1):52-59.
Frisbie, W.P., Youngtae Cho, and R.A. Hummer
  2001 Immigration and the health of Asian and Pacific Islander adults in the United States. *American Journal of Epidemiology* 153(4):372-380.
Frost, F., and K.K. Shy
  1980 Racial differences between linked birth and infant death records in Washington state. *American Journal of Public Health* 70:974-976.
Frost, F., K. Tollestrup, A. Ross, E. Sabotta, and E. Kimball
  1994 Correctness of racial coding of American Indians and Alaska Natives on the Washington state death certificate. *American Journal of Preventive Medicine* 10(5):290-294.
Frye, B.A.
  1995 Use of cultural themes in promoting health among Southeast Asian refugees. *American Journal of Health Promotion* 9(4):269-280.
Geronimus, A.T., and J. Bound
  1998 Use of census-based aggregate variables to proxy for socioeconomic group: Evidence from national samples. *American Journal of Epidemiology* 148(5):475-486.

Gomes, C., and T.G. McGuire
2001 Identifying the sources of racial and ethnic disparities in health care use. Unpublished manuscript.
Gornick, M.E.
2002 Measuring the effects of socioeconomic status on health care. Pp. 45-73 in Institute of Medicine *Guidance for the National Healthcare Disparities Report.* E. Swift, ed., Committee on Guidance for Designing a National Healthcare Disparities Report. Washington, DC: The National Academies Press.
Gornick, M.E., P.W. Eggers, T.W. Reilly, R.M. Mentnech, L.K. Fitterman, L.E. Kucken, and B.C. Vladeck
1996 Effects of race and income on mortality and use of services among Medicare beneficiaries. *New England Journal of Medicine* 335(11):791-799.
Guendelman, S., and B. Abrams
1995 Dietary intake among Mexican-American women: Generational differences and a comparison with white non-Hispanic women. *American Journal of Public Health* 85(1):20-25.
Gustman, A.L., and T.L. Steinmeier
2000 Social Security benefits of immigrants and U.S. born. Chapter 8 in *Issues in the Economics of Immigration,* G.J. Borjas, ed. Chicago, IL: University of Chicago Press.
Hahn, R.A., J. Mulinare, S.M. Teutsch
1992 Inconsistencies in coding of race and ethnicity between birth and death in U.S. infants. *Journal of the American Medical Association* 267(2):259-263.
Harris, D.R.
2002 Racial Classification and the 2000 Census. Prepared for the National Research Council Panel to Review the 2000 Census. Unpublished manuscript.
Hawaii Department of Health
2003 *Health and Vital Statistics: Hawaii Health Statistics 2001.* Available: http://www.hawaii.gov/doh/stats/surveys/hhs/hhs01.html
Henry J. Kaiser Family Foundation
2000 *Health Insurance Coverage in America—1999 Data Update.* Washington, DC: The Kaiser Commission on the Medicaid and the Uninsured.
Hotz, V.J., and J.K. Scholz
2002 Measuring employment and income for low income populations. In *Studies of Welfare Populations: Data Collection and Research Issues,* M. Ver Ploeg, R.A. Moffitt, and C.F. Citro. Panel on Data and Methods for Measuring the Effects of Changes in Social Welfare Programs, Committee on National Statistics, Division of Behavioral and Social Sciences and Education, National Research Council. Washington, DC: National Academy Press.
Institute of Medicine
2003a *Unequal Treatment: Confronting Racial and Ethnic Disparities in Health Care.* B.D. Smedley, A.Y. Stith, and A.R. Nelson, eds. Committee on Understanding and Eliminating Racial and Ethnic Disparities in Health Care, Board on Health Sciences Policy. Washington, DC: The National Academies Press.
2003b *The Future of the Public's Health in the 21st Century.* Committee on Assuring the Health of the Public in the 21st Century. Washington, DC: The National Academies Press.
2002 *Guidance for the National Healthcare Disparities Report.* Elaine K. Smith, ed. Committee on Guidance for Designing a National Healthcare Disparities Report. Washington, DC: The National Academies Press.

Jha, A.K., M.G. Shlipak, W. Hosmer, C.D. Frances, and W.S. Browner
  2001 Racial differences in mortality among men hospitalized in the Veterans Affairs health care system. *Journal of the American Medical Association* 285(3): 297-303.
Kalsbeek, W.
  2003 Sampling minority groups in health surveys. *Statistics in Medicine* 22(9): 1527-1549.
Kaplan, G.A., and J.W. Lynch
  1997 Whither studies on the socioeconomic foundations of population health? (Editorial). *American Journal of Public Health* 87(9):1409-1411.
Karter, A.J., A. Ferrara, J.Y. Liu, H.H. Moffet, L.M. Ackerson, and J.V. Selby
  2002 Ethnic disparities in diabetes complications in an insured population. *Journal of the American Medical Association* 287:2519-2527.
Koroukian, S.M., G.S. Cooper, and A.A. Rimm
  2003 Ability of Medicaid claims data to identify incident cases of breast cancer in the Ohio Medicaid population. *Health Services Research Journal* 38(3):947-960.
Krieger, N.
  1992 Overcoming the absence of socioeconomic data in medical records: Validation and application of a census-based methodology. *American Journal of Public Health* 82:703-710.
Krieger, N., J.T. Chen, P.D. Waterman, D.H. Rehkopf, and S.V. Subramanian
  2003 Race/ethnicity, gender, and monitoring socioeconomic gradients in health: A comparison of area-based socioeconomic measures—The public health disparities geocoding project. *American Journal of Public Health* 93(10):1655-1671.
Lauderdale, D.S., and J. Goldberg
  1996 The expanded racial and ethnic codes in the Medicare data files: Their completeness of coverage and accuracy. *American Journal of Public Health* 86(5):712-716.
Lauderdale, D.S., R.A. Thisted, and J. Goldberg
  1998 Is geographic variation in hip fracture rates related to current/former state of residence? *Epidemiology* 9(5):574-577.
Lewontin, R.C.
  1972 The apportionment of human diversity. Pp. 381-386 in *Evolutionary Biology*. T. Dobzhansky, M.K. Hecht, W.C. Steere, eds. New York: Appleton-Century-Crofts.
Malloy, M.H.
  1998 Effectively delivering the message on infant sleep position. *Journal of the American Medical Association* 280:373-374.
Marin, G., F. Sabogal, B. VanOss, R. Otero-Sabogal, and E. Perez-Stable
  1987 Development of a short acculturation scale for Hispanics. *Hispanic Journal of Behavioral Sciences* 9(2):183-205.
Marmot, M.
  2002 The influence of income on health: View of an epidemiologist. Does money really matter? Or is it a marker for something else? *Health Affairs* 21(2):31-46.
Marmot, M., and M. Wadsworth, eds.
  1997 *Fetal and Early Childhood Environment: Long-term Health Implications*. London, England: Royal Society of Medicine Press Limited.
Massey, D.S.
  2001 Residential segregation and neighborhood conditions in U.S. metropolitan areas. Pp. 391-434 in *America Becoming: Racial Trends and Their Consequences*, N.J. Smelser, W.J. Wilson, and F. Mitchell, eds. Commission on Behavioral and Social Sciences and Education, National Research Council. Washington, DC: National Academy Press.

Massey, D.S., and N. Denton
    1993   *American Apartheid: Segregation and the Making of the Underclass.* Cambridge,
           MA: Harvard University Press.
Mays, V.M., N. Ponce, D.L. Washington, and S.D. Cochran
    2003   Classification of race and ethnicity: Implications for public health. *Annual Review
           of Public Health* 24:83-110.
Moore, J.C., L.L. Stinson, and E.J. Welniak, Jr.
    1997   *Income Measurement Error in Surveys: A Review.* (Statistical Research Report).
           Washington, DC: U.S. Census Bureau.
National Center for Health Statistics
    2003   *Health, United States, 2003 with Chartbook on Trends in the Health of Americans.*
           Hyattsville, MD: National Center for Health Statistics.
    2002   Deaths: Final data for 2000. National Vital Statistics Reports 50(15):September 16.
           Available: http://www.cdc.gov/nchs/pressroom/02facts/final2000.htm (June 2004).
    2000   *Report of the Panel to Evaluate the U.S. Standard Certificates.* Washington, DC:
           Division of Vital Statistics, National Center for Health Statistics. Available: http://
           www.cdc.gov/nchs/data/dvs/panelreport_acc.pdf (September 29, 2003).
National Governors Association
    2003   *Maternal and Child Health Update 2002: State Health Coverage for Low-Income
           Pregnant Women, Children, and Parents.* Washington, DC: Health Policy Division.
National Health Law Program
    in   *Assessment of State Laws, Regulations, and Practices Affecting the Collection and
    press Reporting of Racial and Ethnic Data by Health Insurers and Managed Care Plans.*
           Prepared for the DHHS Office of Minority Health.
National Institutes of Health
    1992   *A Mortality Study of 1.3 Million Persons by Demographic, Social, and Economic
           Factors: 1979-1985 Follow-Up.* Bethesda, MD: National Institutes of Health.
National Research Council
    2004   *Measuring Racial Discrimination.* R.M. Blank, M. Dabady, and C.F. Citro, Eds.
           Panel on Methods for Assessing Discrimination, Committee on National Statistics,
           Division of Behavioral and Social Sciences and Education. Washington, DC: The
           National Academies Press.
    2003   *Improving Racial and Ethnic Data in Health: Report of a Workshop.* Dan Melnick,
           ed. Committee on National Statistics, Division of Behavioral and Social Sciences
           and Education. Washington, DC: The National Academies Press.
    2000   *Improving Access to and Confidentiality of Research Data: Report of a Workshop.*
           C. Mackie and N. Bradburn, eds. Committee on National Statistics, Commission
           on Behavioral and Social Sciences and Education. Washington, DC: National Acad-
           emy Press.
    1993   *Private Lives and Public Policies: Confidentiality and Accessibility of Government
           Statistics.* G.T. Duncan, T.B. Jabine, and V.A. de Wolf, eds. Panel on Confidential-
           ity and Data Access, Committee on National Statistics, Commission on Behavioral
           and Social Sciences and Education. Washington, DC: National Academy Press.
Oaks, J.M., and P.H. Rossi
    2003   The measurement of SES in health research: Current practice and steps toward a
           new approach. *Social Sciences and Medicine* 56(4):769-784.
Office of Management and Budget
    2000a *Guidance on Aggregation and Allocation of Data on Race for Use in Civil Rights
           Monitoring and Enforcement.* (OMB Bulletin Number 00-02). Washington, DC:
           Office of Management and Budget.

2000b   *Provisional Guidance on the Implementation of the 1997 Standards for Federal Data on Race and Ethnicity.* Washington, DC: Office of Management and Budget. Available: http://www.whitehouse.gov/omb/inforeg/re_guidance2000update.pdf (August 27, 2003).

1997   Revisions to the standards for the classification of federal data on race and ethnicity. *Federal Register Notice,* October 30.

1977   *Directive 15: Race and Ethnic Standards for Federal Statistics and Administrative Reporting.* Washington, DC: Office of Management and Budget.

Oliver, M.L., and T.M. Shapiro

1995   *Black Wealth, White Wealth: A New Perspective on Racial Inequality.* New York: Routledge.

Omi, M., and H. Winant

1986   *Racial Formation in the United States: From the 1960s to the 1980s.* New York: Routledge.

Parker, J.D., N. Schenker, D.D. Ingram, J.A. Weed, K.E. Heck, and J.H. Madans

2003   *Bridging Between Two Standards for Collecting Information on Race and Ethnicity; An Application to Census 2000 and Vital Rates.* Hyattsville, MD: National Center for Health Statistics.

Pear, R.

2004   Taking spin out of report that made bad into good. *New York Times,* February 22.

Perez-Stable, E.J., A. Napoles-Springer, and J.M. Miramontes

1997   The effects of ethnicity and language on medical outcomes of patients with hypertension or diabetes. *Medical Care* 35(12):1212-1219.

Perot, R.T., and M. Youdelman

2001   *Racial, Ethnic, and Primary Language Data Collection in the Health Care System: An Assessment of Federal Policies and Practices.* Washington, DC: The Commonwealth Fund.

Peterson E.D., S.M. Wright, J. Daley, and G.E. Thibault

1994   Racial variation in cardiac procedure use and survival following acute myocardial infarction in the Department of Veterans Affairs. *Journal of the American Medical Association* 272(15):455-461.

Ricketts, T.C.

2002   Geography and disparities in health care. In Institute of Medicine, *Guidance for the National Healthcare Disparities Report,* E.K. Swift, ed. Washington, DC: The National Academies Press.

Rodgers, W.L., C. Brown, and G.J. Duncan

1993   Errors in survey reports of earnings, hours worked, and hourly wages. *Journal of the American Statistical Association* 88(December):1208-1218.

Roemer, M.

2000   Reconciling March CPS Money Income with the National Income and Product Accounts: An Evaluation of CPS Quality. Unpublished paper, Income Statistics Branch, Bureau of the Census, August 10.

1999   Assessing the quality of the March Current Population Survey and the Survey of Income and Program Participation Income Estimates, 1990-1996. Unpublished paper, Income Statistics Branch, Bureau of the Census, June 16.

Sandefur, G.D., M. Martin, J. Eggerling-Boeck, S.E. Mannon, and A.A. Meier

2002   An Overview of Racial and Ethnic Demographic Trends. Pp. 40-102 in *America Becoming: Racial Trends and Their Consequences,* N.J. Smelser, W.J. Wilson, and F. Mitchell (eds.). Division of Behavioral and Social Sciences and Education. Washington, DC: National Academy Press.

Schneider, E.C., A.M. Zaslavsky, and A.M. Epstein
    2002   Racial disparities in the quality of care for enrollees in Medicare Managed Care. *Journal of the American Medical Association* 287(10):1288-1294.
Scott, C.G.
    1999   Identifying the race or ethnicity of SSI recipients. *Social Security Bulletin* 62(4):9-20.
Singh, G.K., B.A. Miller, B.F. Hankey, and B.K. Edwards
    2003   *Area Socioeconomic Variations in U.S. Cancer Incidence, Mortality, Stage, Treatment, and Survival, 1975-1999.* NCI Cancer Surveillance Monograph Series, Number 4. NIH Publication No. 03-5417. Bethesda, MD: National Cancer Institute.
Skinner, J., J. Weinstein, S. Sporer, and J. Wennberg
    2003   Racial, ethnic, and geographic disparities in rates of knee arthroplasty among Medicare patients. *New England Journal of Medicine* 349(14):1350-1359.
Smith, D.B.
    1999   *Health Care Divided: Race and Healing a Nation.* Ann Arbor: University of Michigan Press.
Smith, J.
    1999   Healthy bodies and thick wallets: The dual relation between health and economic status. *Journal of Economic Perspectives* 13(20):145-166.
Smyser, M., J. Krieger, and D. Solet
    1999   *The King County Ethnicity and Health Survey.* Special report prepared by The Epidemiology, Planning and Evaluation Unit, Public Health—Seattle and King County. Seattle: Public Health Seattle and King County.
Sorlie, P.D., E. Backlund, and J.B. Keller
    1995   U.S. mortality by economic, demographic, and social characteristics: The National Longitudinal Mortality Study. *American Journal of Public Health* 85(7):949-956.
Sorlie, P.D., E. Rogot, and N.J. Johnson
    1992   Validity of demographic characteristics on the death certificate. *Epidemiology* 3(2):181-184.
Substance Abuse and Mental Health Services Administration
    2002   *Drug Abuse Warning Network: Development of a New Design, Methodology Report.* Washington, DC: U.S. Department of Health and Human Services.
Sundquist, J., and M. Winkleby
    2000   Country of birth, acculturation status and abdominal obesity in a national sample of Mexican-American women and men. *International Journal of Epidemiology* 29(3):470-477.
Swanson, G.M., A.G. Schwartz, and R.W. Burrows
    1984   An assessment of occupation and industry data from death certificates and hospital medical records for population-based cancer surveillance. *American Journal of Public Health* 74(1984):464-467.
Taylor, J.A., and C.F. Liu
    2003   The Potential of Record Linkage for Health Research and Program Evaluation. Presentation at the Annual Meetings of the National Association of Health Data Organizations, December, Atlanta, GA.
Turrell, G., B. Olderburg, I. McGuffog, and R. Dent
    1999   *Socioeconomic Determinants of Health: Towards a National Research Program and a Policy Intervention Agenda.* Canberra, Australia: Queensland University of Technology, School of Public Health, Ausinfo.

Unger, J.B., T.B.Cruz, L.A. Rohrbach, K.M. Ribisl, L. Baezconde-Garbanati, X. Chen, D.R. Trinidad, and C.A. Johnson
  2000 English language use as a risk factor for smoking initiation among Hispanic and Asian American adolescents: Evidence for mediation by tobacco-related beliefs and social norms. *Health Psychology* 19(5):403-410.
U.S. Bureau of the Census
  1996 *Findings on Questions on Race and Hispanic Origin Tested in the 1996 National Content Survey.* (Population Division Working Paper No. 16). Washington, DC: U.S. Government Printing Office.
  1993 *U.S. Census of Population 1990, Social and Economic Characteristics.* Washington, DC: U.S. Government Printing Office.
U.S. Department of Commerce
  2001a *Poverty in the United States: 2000.* Current Population Reports, Consumer Income. Washington, DC: U.S. Government Printing Office.
  2001b *Money Income in the United States: 2000.* Current Population Reports, Consumer Income. Washington, DC: U.S. Government Printing Office.
U.S. Department of Health and Human Services
  2003a *National Healthcare Disparities Report.* Rockville, MD: U.S. Department of Health and Human Services, Agency for Healthcare Research and Quality.
  2003b *Directory of Health and Human Services Data Resources.* Office of the Assistant Secretary for Planning and Evaluation. Available: http://aspe.hhs.gov/datacncl/datadir/index.shtml (September 29, 2003).
  2002 *45 CFR Part 162. Title 45: Public Welfare. Subtitle A: Department of Health and Human Services. Part 162: Administrative Requirements.* Available: http://www.access.gpo.gov/nara/cfr/waisidx_02/45cfr162_02.html
  2000 *Healthy People 2010: Understanding and Improving Health.* Washington, DC: U.S. Government Printing Office.
  1999 *Improving the Collection and Use of Racial and Ethnic Data in HHS.* Joint report of the HHS Data Council Working Group on Racial and Ethnic Data and the Data Work Group of the HHS Initiative to Eliminate Racial and Ethnic Disparities in Health. Washington, DC: U.S. Department of Health and Human Services.
  1997 *U.S. Vital Statistics System: Major Activities and Developments, 1950-1995.* National Center for Health Statistics, Centers for Disease Control and Prevention. Hyattsville, MD: U.S. Department of Health and Human Services.
  1990 *Healthy People 2000.* Washington, DC: U.S. Department of Health and Human Services, Office of Disease Prevention and Health Promotion.
  1985 *Report of the Secretary's Task Force on Black and Minority Health.* Washington, DC: U.S. Department of Health and Human Services.
U.S. Public Health Service
  1993 *One Voice, One Vision—Recommendations to the Surgeon General to Improve Hispanic/Latino Health.* (Surgeon General's National Hispanic/Latino Health Initiative). Washington, DC: Office of the Surgeon General.
  1992 *Improving Minority Health Statistics.* (Report of the PHS Task Force on Minority Health Data). Washington, DC: U.S. Government Printing Office.
van Rossum, C.T.M., M.J. Shipley, H. van de Mheen, D.E. Grobbee, and M.G. Marmot
  2000 Employment grade differences in cause specific mortality: A 25 year follow up of civil servants from the first Whitehall study. *Journal of Epidemiology and Community Health* (March 2000) 54:178-184.
Vega, W.A., and H. Amaro
  1994 Latino outlook: Good health, uncertain prognosis. *Annual Review of Public Health* 15:39-67.

Venti, S.F., and D.A. Wise
   1999   Lifetime earnings, saving choices, and wealth at retirement. In *Wealth, Work, and Health: Innovations in Survey Measurement in the Social Sciences*, J. Smith and R. Willis, eds. Ann Arbor: University of Michigan Press.

Waksberg, J., D. Levine, and D. Marker
   2000   *Assessment of Major Federal Data Sets for Analyses of Hispanic and Asian or Pacific Islander Subgroups and Native Americans*, Task 2 Report: Inventory of Selected Existing Federal Databases. Rockville, MD: Westat.

Wheaton, L., and L. Giannarelli
   2000   Underreporting of means-tested transfer programs in the March CPS. In *American Statistical Association 2000 Proceedings of the Section on Government Statistics and Section on Social Statistics*. Alexandria, VA: American Statistical Association.

Williams, D.R.
   2002   Racial/ethnic variations in women's health: The social embeddedness of health. *American Journal of Public Health* 92(4):588-597.
   1996   Race/ethnicity and socioeconomic status: Measurement and methodological issues. *International Journal of Health Services* 26(3):483-505.

Williams, D.R., and C. Collins
   1995   U.S. socioeconomic and racial differences in health: Patterns and explanations. *Annual Review of Sociology* 21:349-386.

Wilson, W.J.
   1987   *Truly Disadvantaged: The Inner City, the Underclass, and Public Policy*. Chicago: University of Chicago Press.

Winslow, R.
   2003   Aetna is collecting racial data to monitor medical disparities. *Wall Street Journal*, March 5, A1.

Wu, S.
   2003   The effects of health events on the economic status of married couples. *Journal of Human Resources* 38(1):209-230.

Yeh, C.J.
   2003   Age, acculturation, cultural adjustment, and mental health symptoms of Chinese, Korean, and Japanese immigrant youths. *Cultural Diversity and Ethnic Minority Psychology* 9(1):34-48.

Yi, J.K.
   1995   Acculturation, access to care, and use of preventive health services by Vietnamese women. *Asian American and Pacific Islander Journal of Health* 3(1):30-41.

Youdelman, M.
   2002   The legal environment for collecting racial and ethnic data in health and health care. Comments prepared for the Workshop on Improving the Collection of Race and Ethnicity Data in Health, December 13-14, Washington, DC.

Yu, S.M., Z.J. Huang, R.H. Schwalber, M.D. Overpeck, and M.D. Kogan
   2002   Association of language spoken at home with health and school issues among Asian American adolescents. *Journal of School Health* 72(5):192-198.

Zaslavsky, A.M., J.N. Hochheimer, E.C. Schneider, P.D. Cleary, J.J. Seidman, E.A. McGlynn, J.W. Thompson, C. Sennett, and A.M. Epstein
   2000   Impact of sociodemographic case mix on the HEDIS measures of health plan quality. *Medical Care* 38:981-992.

Appendix A

Descriptions of National Health and Health Care Surveys

NATIONAL HOUSEHOLD SURVEYS ON HEALTH-RELATED TOPICS

Behavioral Risk Factor Surveillance System (BRFSS)

Agency: National Center for Chronic Disease Prevention and Health Promotion, Centers for Disease Control and Prevention

Main Purpose of Data Collection: To collect state-level data on actual risk behaviors, as opposed to attitudes or feelings. The data are used to track and reduce health risk behaviors and associated illnesses.

Periodicity of Data Collection: Monthly since 1984

Type of Survey:
- Cross-sectional
- Telephone interview

Sample Size: 1,200-7,000 interviewed yearly in each state; 100-700 interviewed monthly in each state.

Racial and Ethnic Data Collected: Race: American Indian/Alaska Native; Asian or Pacific Islander; Black;

White; Other; Refused Specification; Don't Know/Not Sure

Ethnicity: Hispanic Origin; Not Hispanic Origin; Don't Know; Refused Specification

SEP Data Collected: Education, employment, income

Language and Acculturation Data Collected: Not collected

Consumer Assessment of Health Plans Survey (CAHPS)

Agency: Center for Beneficiary Choices, Centers for Medicare and Medicaid Services

Main Purpose of Data Collection: To collect information on consumers' assessments of health plans in order to provide other consumers with easily accessible and understandable data to compare alternative health care providers.

Periodicity of Data Collection: 1995, ongoing

Type of Survey:
• Cross-sectional
• Kit of survey and report tools provided to health care providers. Information gathered is then reported to consumers.

Sample Size: 2001: 780,000 individuals completed the survey

Racial and Ethnic Data Collected: Follows 1997 revised OMB standards

SEP Data Collected: Highest grade level completed (these are core questions; other data may be collected specific to the type of services provided)

Language and Acculturation Data Collected: The survey is offered in English and Spanish.

Medical Expenditures Panel Survey (MEPS)

Agency:	Sponsoring organizations: Agency for Healthcare Research and Quality (AHRQ) and National Center for Health Statistics (NCHS)
Main Purpose of Data Collection:	To provide data about health care use and costs to improve economic projections. There are three main components—Household (HC), Insurance (IC), and Medical Provider (MPC); and one supplement—Nursing Home Component (NHC).
Periodicity of Data Collection:	Annual. HC and IC conducted from 1996-2000. NHC was a one-time survey in 1996. MPC is ongoing.
Type of Survey:	• Longitudinal: yearly sample followed for 2 years with five in-person interviews over the 2 years (approximately every 5-6 months). • HC is conducted as a personal interview in respondents' households. Both the IC and MPC consist of telephone interviews and mailed survey materials. The NHC was a year-long survey that consisted of in-person interviews with nursing home sources and telephone interviews with next of kin. • MPC is not available for public use.
Sample Size:	HC: 1997: 12,600 households, 34,000 individuals 1998: 10,500 households, 23,000 individuals IC: Each year approximately 36,900 establishments and individuals (7,000 establishments from HC, 27,000 establishments from the Census Bureau's list of private-sector businesses, 1,900 government offices from Census list of government employers,

and 1,000 individuals from IRS list of self-employed)

MPC: Approximately 22,600 providers (2,700 hospitals, 12,400 office-based physicians, 7,000 separate-billing hospital physicians, 500 home health providers)

Racial and Ethnic Data Collected: HC: Race: American Indian, Aleut; Asian; Black; White; Other Ethnicity: Hispanic or Not Hispanic

NHC: Race: American Indian or Alaska Native; Asian or Pacific Islander; Black; White; Other Ethnicity: Hispanic or Not Hispanic

IC: Not collected

MPC: Not collected

SEP Data Collected: HC: education, employment, income/poverty status, health care expenditures, and wealth (assets/debts)

NHC: income and insurance coverage

IC: Not collected

MPC: Not collected

Language and Acculturation Data Collected: HC: Language of interview

IC, MPC, NHC: Not collected

Medicare Current Beneficiary Survey (MCBS)

Agency: Office of Research, Development, and Information (ORDI), Centers for Medicare and Medicaid Services

Main Purpose of Data Collection: To collect data on Medicare beneficiaries' care received, cost of care, and source of payment to aid Health Care Financing Administration's oversight of Medicare and Medicaid programs.

Periodicity of Data Collection:	Annually since 1991
Type of Survey:	• Longitudinal. A rotating panel is interviewed three times a year. Can be used as a time series. • In-person interview using CAPI
Sample Size:	Approximately 12,000 each year
Racial and Ethnic Data Collected:	Through 1997: Race: White; Black; American Indian or Alaska Native; Asian or Pacific Islander; Other Ethnicity: Hispanic or Not Hispanic Beginning in 1998: Race and Ethnicity: Follows 1997 revised OMB standards
SEP Data Collected:	Income, assets, family supports, quality of life, education
Language and Acculturation Data Collected:	Not collected

National Health and Nutrition Examination Survey (NHANES)

Agency:	National Center for Health Statistics, Centers for Disease Control and Prevention

NHANES I

Main Purpose of Data Collection:	To collect and disseminate health and nutrition information and statistics. Also to measure and monitor the nutritional status of adults and children.
Periodicity of Data Collection:	1971-1974 and 1974-1975
Type of Survey:	• Cross-sectional • Interview and physical examination at mobile exam center (including vision and hearing, cardiovascular fitness and muscle strength, leg circu-

lation and foot sensation tests; body fat, height, and weight measurements; dietary interview; body composition scan; dental check; and laboratory work)

Sample Size:	1971-1974	1974-1975
Sample Size	28,043	4,288
Interviewed	27,753	4,220
Examined	30,749	3,059

Racial and Ethnic Data Collected: Race (1971-1974): White; Negro

Origin or Descent (1971-1974): German; Irish; Italian; French; Polish; Russian; English; Spanish; Mexican; Chinese; Japanese; American Indian; and Another group not listed

Race (1974-1975): White; Negro

Origin or Descent (1974-1975): German; Irish; Italian; French; Polish; Russian; English; Welsh; Mexican; Mexican-American; Chicano; Mexicano; Puerto Rican; Cuban; Central or South American; Other Spanish; Negro; Black

SEP Data Collected: Income, education, poverty status, employment (industry)

Language and Acculturation Data Collected: Language spoken at home

NHANES II

Main Purpose of Data Collection: To continue data collection from NHANES I on health and nutritional status of the population.

Periodicity of Data Collection: 1976-1980

Type of Survey: See NHANES I

Sample Size: Sample 27,801
Interviewed 25,286
Examined 20,322

Racial and Ethnic Data Collected: Race: White; Black; Other

Ethnicity: Countries of Central or South America; Chicano; Cuban; Mexican; Mexicano; Mexican American; Puerto Rican; Other Spanish; Other European (such as German, Irish, English, or French); Black/Negro/African American; American Indian or Alaska Native; Asian or Pacific Islander (such as Chinese, Japanese, Korean, Filipino, Samoan); Other

SEP Data Collected: Poverty status, income, education, employment (industry, occupation)

Language and Acculturation
  Data Collected: Language spoken at home

NHANES III

Main Purpose of Data Collection: To continue data collection from NHANES I and II on health and nutritional status of the population.

Periodicity of Data Collection: 1988-1994

Type of Survey: • Cross-sectional
                                        • Interview at respondent's home (using pencil-and-paper interviewing [PAPI] during Phase I and computer-assisted personal interview [CAPI] during Phase II) and physical examination

Sample Size: Phase I: 20,3000 (17,500 interviewed, 15,600 examined)

                                        Phase II: 19,400 (16,500 interviewed, 15,200 examined)

Racial and Ethnic Data Collected: Race: Aleut/Eskimo/American Indian; Asian or Pacific Islander; Black; White; Other

Ethnicity: Mexican/Mexican Ameri-

can; Other Latin American or Other
Spanish; Not of Hispanic Origin

SEP Data Collected: Education, employment, income/
poverty status, health care expendi-
tures

*Language and Acculturation
  Data Collected:*

Language spoken primarily at home,
work, and school; language in which
the interview was conducted (personal
and examination); respondent, maternal, and paternal nativity

1999-Current NHANES

Main Purpose of Data Collection: To continue data collection from
NHANES I, II, and III on health and
nutritional status of the population.

Periodicity of Data Collection: 1999-2003 (the latest release is 1999-
2000 data)

Type of Survey: Cross-sectional

Sample Size: 1999-2000 data release: 9,965 persons

Racial and Ethnic Data Collected: Follows 1997 revised OMB standards

SEP Data Collected: Education, employment (industry,
occupation, job duties), income

*Language and Acculturation
  Data Collected:*

Respondent, paternal, and maternal
nativity; language spoken mainly at
home, as a child, and with friends;
languages respondent can read and
speak; language in which the respondent thinks

National Health Interview Survey (NHIS)

Agency: National Center for Health Statistics,
Centers for Disease Control and
Prevention

Main Purpose of Data Collection:	To provide national-level general health statistics to monitor the health status of the U.S. population.
Periodicity of Data Collection:	Continuously since 1957 (new samples are drawn on a weekly basis). The content of the survey has been updated every 10-15 years.
Type of survey:	• Cross-sectional • Surveys conducted in respondents' homes
Sample Size:	2002 Survey: Family Core: 93,138 individuals Sample Adult Core: 31,044 individuals Sample Child Core: 12,524 individuals Total:136,706
Racial and Ethnic Data Collected:	Race: For public use: 1997 revised OMB standards
	Additional racial categories not available for public use: Native Hawaiian; Guamanian; Samoan; Other Pacific Islander; Asian Indian; Chinese; Filipino; Japanese; Korean; Vietnamese
	Ethnicity: Hispanic, Central or South American; Other Latin American; Other Spanish; Hispanic/Latino/Spanish type unknown; Not Hispanic/Spanish origin
SEP Data Collected:	Employment, family income, education
Language and Acculturation Data Collected:	Place of birth, citizenship status

National Immunization Survey (NIS)

Agency:	National Center for Health Statistics, Centers for Disease Control and Prevention

Main Purpose of Data Collection:	To provide vaccination data for public health partners.
Periodicity of Data Collection:	Quarterly, 1995-2001
Type of Survey:	• Cross-sectional • Telephone survey; respondents' immunization information verified by vaccination providers
Sample Size:	About 30,000 infants and children between the ages of 12 and 35 months
Racial and Ethnic Data Collected:	Race and ethnicity of mother and child collected: Race: White; Black; American Indian; Asian; Other; Don't Know; Refused Specification Ethnicity: Not Spanish/Hispanic; Mexican; Mexican American; Chicano; Puerto Rican; Cuban; Other Spanish; Don't Know; Refused Specification
SEP Data Collected:	Maternal education, poverty status, income
Language and Acculturation Data Collected:	Language of interview (English or Spanish)

National Longitudinal Study of Adolescent Health (AddHEALTH)

Agency:	Demographic and Behavioral Sciences Branch, National Institute for Child Health and Human Development, National Institutes of Health
Main Purpose of Data Collection:	To provide a comprehensive view of the health and health behaviors of adolescents and the antecedents—personal, interpersonal, familial, and environmental—of these outcomes.
Periodicity of Data Collection:	September 1994-December 1995;

	April-August 1996; August 2001- April 2002
Type of Survey:	Longitudinal
Sample Size:	21,000 adolescents were included in the original sample; 15,000 young adults completed the most recent interview.
Racial and Ethnic Data Collected:	First question about race/ethnicity asks respondents if they are of Hispanic or Latino origin. The next question offers the following choices: white; black or African American; American Indian or Native American; Asian or Pacific Islander; other. Respondents are allowed to choose as many races as they wish.
SEP Data Collected:	In the second wave of interviews, respondents were asked how much they typically work and earn during the school year and over the summer.
Language and Acculturation Data Collected:	Duration of domicile in the United States is collected. The respondent is asked what language is usually spoken in the home. Responses include English, Spanish, Other (please record language), and don't know. If respondents say they are Hispanic or Asian, they are asked to choose from a list of countries to clarify their background.

National Maternal and Infant Health Survey (NMIHS)

Agency:	National Center for Health Statistics, Centers for Disease Control and Prevention
Main Purpose of Data Collection:	To provide data to researchers on factors affecting adverse outcomes of pregnancy.

Periodicity of Data Collection:	1988 with 1991 follow-up on the live births (closed)
Type of Survey:	• Cross-sectional • Interviews with mothers, and vital records information obtained from hospitals
Sample Size:	9,935 live births, 3,309 fetal deaths, and 5,332 infant deaths
Racial and Ethnic Data Collected:	Race and ethnicity of mother and father: Survey Data Collection: Race: American Indian/Alaskan; Asian or Pacific Islander; Black; White Ethnicity: Follows 1997 revised OMB standards Vital Records Data: Race: See above Ethnicity: Non-Spanish; Puerto Rican; Cuban; Mexican; Central or South American; Other, unknown Spanish; Not classifiable
SEP Data Collected:	From survey data: Maternal/paternal education/vocational training, employment
Language and Acculturation Data Collected:	From vital records: Maternal nativity

National Mortality Followback Survey (NMFS)

Agency:	National Center for Health Statistics, Centers for Disease Control and Prevention
Main Purpose of Data Collection:	To provide data to analyze the causes of disease, and other issues related to health and mortality.
Periodicity of Data Collection:	Survey years: 1961, 1962-1963,

1964-1965, 1966-1968, 1986, 1993
(closed)

Type of Survey:

- Cross-sectional
- Collect data from death certificates and next-of-kin interviews

Sample Size:

1993: 22,957 death certificates (including 9,636 death certificates selected with certainty)

Racial and Ethnic Data Collected:

Race: White; Black; American Indian; Eskimo; Aleut; Chinese; Filipino; Hawaiian; Korean; Vietnamese; Japanese; Asian Indian; Samoan; Guamanian; Other Asian/Pacific Islander; Other Race

Ethnicity: Puerto Rican; Cuban; Mexican/Mexicano; Mexican American; Chicano; Other Latin American; Other Spanish

SEP Data Collected:

Medical expenditures, education, employment, income and assets

Language and Acculturation Data Collected:

Country of origin

National Survey on Drug Use and Health (NSDUH)

Agency:

Office of Applied Studies, Substance Abuse and Mental Health Services Administration

Main Purpose of Data Collection:

To collect data for the study of patterns of substance use. It provides the primary data source of illegal drug use. Data are also collected on mental health and on tobacco and alcohol use and abuse.

Periodicity of Data Collection:

Annual since 1971

Type of Survey:

- Cross-sectional
- Interviewer-administered and self-

	administered using computer-assisted interviewing (CAI)
Sample Size:	Approximately 70,000 (12 years and older) surveyed each year. Sample size varies year to year.
Racial and Ethnic Data Collected:	Race: White; Black/African American; American Indian or Alaska Native; Asian or Pacific Islander; Other
	Ethnicity: Hispanic or Spanish Origin or Descent; Not of Hispanic or Spanish Origin or Descent
SEP Data Collected:	Education, employment, family income
Language and Acculturation Data Collected:	Respondents are asked what country they were born in and how long they have lived in the United States.

National Survey of Family Growth (NSFG)

Agency:	National Center for Health Statistics, Centers for Disease Control and Prevention
Main Purpose of Data Collection:	To provide national data on the health of women and infants as well as on marriage, divorce, contraception, and infertility.
Periodicity of Data Collection:	Year (Cycle) 1973 (1), 1976 (2), 1988 (3), 1995 (4), 2002 (5), 2003 (6)
Type of Survey:	• Cross-sectional • In person, at respondents' homes. The first section is completed using CAPI (computer-assisted personal interviewing), and the last section is entered by the respondent using audio computer-assisted self-interviewing. • Interview only women ages 15-44. Cycles 1 and 2 only include women

who had ever been married or had
their own children living with them.
Cycle 6 includes men.

Sample Size:

Cycle	Sample
1	9,797
2	8,611
3	7,969
4	8,450
5	10,847
6	Not available

Racial and Ethnic Data Collected: Race/Ethnicity: American Indian/
Alaskan Native; Asian or Pacific
Islander; Black not of Hispanic origin;
White not of Hispanic origin; His-
panic

SEP Data Collected: Education/vocational training, employment

Language and Acculturation
Data Collected: Not collected

Youth Risk Behavior Surveillance System (YRBSS)

Agency: National Center for Chronic Disease
Prevention and Health Promotion,
Centers for Disease Control and
Prevention

Main Purpose of Data Collection: To provide data for the study of the
prevalence of health risk behaviors of
young people.

Periodicity of Data Collection: Biennial since 1990

Type of survey: • Cross-sectional
• The local school-based survey (of
9th through 12th grade students) is a
self-administered questionnaire.
• In 1992 there was a household
survey, and in 1995 there was a
national college survey to supplement
the school surveys.

Sample Size: Surveys 9th-12th graders. The average

sample size for state and local surveys: 1,819 students. The national
survey in 2001 had 13,000 usable
questionnaires.

Racial and Ethnic Data Collected: Follows 1997 revised OMB standards

SEP Data Collected: Not collected

Language and Acculturation
  Data Collected: Not collected

HEALTH CARE ESTABLISHMENT-BASED
SURVEY DATA COLLECTIONS

(in alphabetical order)

Healthcare Cost and Utilization Project (HCUP)

Agency: Center for Organization and Delivery
                               Studies, Agency for Healthcare Re-
                               search and Quality

Main Purpose of Data Collection: To bring together the data collection
                               efforts of state data organizations,
                               hospital associations, private data
                               organizations, and the federal govern-
                               ment to create a national information
                               resource of health care data.

Periodicity of Data Collection: Ongoing since 1988

How Data are Collected: From various state databases

Unit of Analysis: Hospital discharge

Sample Size: HCUP databases include vast
                               amounts of data. For example, one of
                               the databases, the Nationwide Inpa-
                               tient Sample, includes data on ap-
                               proximately 7 million hospital stays.

Racial and Ethnic Data Collected: Because of differences in the coding of
                               race and ethnicity across the state
                               data systems that provide data to
                               HCUP, the following racial and ethnic
                               categories have been employed: (1)
                               White, (2) Black, (3) Hispanic, (4)

Asian or Pacific Islander, (5) Native American, (6) Other. About half of the states supplying data for HCUP provide complete reporting of race/ethnicity in their data.

SEP Data Collected: Median income for Zip Code

Language and Acculturation
*   Data Collected:*   Not collected

National Ambulatory Medical Care Survey (NAMCS)

Agency: National Center for Health Statistics, Centers for Disease Control and Prevention

Main Purpose of Data Collection: To collect data on ambulatory care visits made to physician offices in the United States.

Periodicity of Data Collection: Conducted annually 1973-1981, again in 1985, and annually since 1989

How Data are Collected: A national sample of nonfederally employed, office-based physicians are each randomly assigned a one-week reporting period. During that week, data for a systematic random sample of visits are recorded by the physician or office staff.

Unit of Analysis: Visit

Sample Size: 27,369 visits in the 2000 survey

Racial and Ethnic Data Collected: Uses OMB standards, but does not collect more than one race. The physician or office staff answers the race/ethnicity question, so the response is based on the physician's knowledge of the patient or on observation.

SEP Data Collected: Not collected

Language and Acculturation
  Data Collected: Not collected

National Home and Hospice Care Survey (NHHCS)

Agency: National Center for Health Statistics,
                                         Centers for Disease Control and
                                         Prevention

Main Purpose of Data Collection: To collect data from home and hos-
                                         pice care agencies in the United
                                         States.

Periodicity of Data Collection: 1992, 1993, 1994, 1996, 1998, and
                                         2000

How Data are Collected: The survey uses a two-stage probabil-
                                         ity sampling design. In the first stage,
                                         a stratified sample of facilities was
                                         taken; facilities were stratified by type
                                         (home health agencies, hospices, and
                                         mixed agencies), Metropolitan Statis-
                                         tical Area, region, and certification
                                         status. In the second stage, lists of
                                         current residents and discharges were
                                         constructed for each agency so that
                                         six current residents and six dis-
                                         charges could be selected using a
                                         systematic probability sample. In-
                                         cluded discharges that occurred be-
                                         cause of the death of the patient.

Unit of Analysis: Agency and individual

Sample Size: 1,425 facilities, current residents, and
                                         discharges in the 2000 survey

Racial and Ethnic Data Collected: Uses OMB racial standards, but
                                         includes Don't Know. Hispanic/
                                         Latino origin was collected as Yes,
                                         No, Don't Know.

SEP Data Collected: Not collected

Language and Acculturation
  Data Collected: Not collected

National Hospital Discharge Survey (NHDS)

Agency:

National Center for Health Statistics, Centers for Disease Control and Prevention

Main Purpose of Data Collection:

To provide information annually on the inpatient use of hospitals in the United States.

Periodicity of Data Collection:

Annually since 1965

How Data are Collected:

Information on diagnoses, surgical and nonsurgical procedures, and patient characteristics is abstracted from the sample of medical records.

Unit of Analysis:

Hospital discharge

Sample Size:

Approximately 270,000 discharge records from approximately 500 nonfederal, short-stay hospitals each year

Racial and Ethnic Data Collected:

American Indian/Eskimo/Aleut; Asian or Pacific Islander; Black; White; Other—Specify; and Not Stated. Ethnic categories: Hispanic Origin; Non-Hispanic; Not Stated

SEP Data Collected:

Not collected

Language and Acculturation Data Collected:

Not collected

National Nursing Home Survey (NNHS)

Agency:

National Center for Health Statistics, Centers for Disease Control and Prevention

Main Purpose of Data Collection:

To collect data from nursing homes, their residents, and their staff.

Periodicity of Data Collection:

1973-74, 1977, 1985, 1995, 1997, and 1999

How Data are Collected:

In the first stage, facilities were selected from a sample frame consisting of nursing homes with 3 or more beds, those certified by Medicare or Medicaid, and those with a state license to operate as a nursing home. Facilities were stratified by number of beds and certification status, and nursing homes were selected using systematic probability sample. In the second stage, lists of current residents and discharges were constructed for each facility, so that six current residents and six discharges could be selected using a systematic probability sample. Discharges that occurred because of death are included in the sample.

Unit of Analysis:

Care providers, care recipients, and facilities

Sample Size:

1,423 facilities, 8,215 current residents, and 6,913 discharges

Racial and Ethnic Data Collected:

Uses OMB standards, but includes Don't Know. Hispanic/Latino origin was collected as Yes, No, Don't Know.

SEP Data Collected:

Not collected

Language and Acculturation Data Collected:

Not collected

National Survey of Substance Abuse Treatment Services (N-SSATS)

Agency:

Substance Abuse and Mental Health Services Administration

Main Purpose of Data Collection:

To collect the data on location, characteristics, services offered, and utilization for all facilities in the Inventory of Substance Abuse Treatment Services (I-SATS), which is a listing of all the known public and private

	substance abuse treatment facilities in the United States and its territories.
Periodicity of Data Collection:	1976-1980, 1982, 1984, 1987, 1989-1993, 1995-1998, 1999 (abbreviated version because of redesign), 2000, 2002. The survey has been known by several different names.
How Data are Collected:	There is a full mail survey with a telephone follow-up in which a facility representative fills out the form.
Unit of Analysis:	Facility
Sample Size:	14,622 facilities eligible in 2000, and a 94 percent response rate
Racial and Ethnic Data Collected:	Not collected
SEP Data Collected:	Not collected
Language and Acculturation Data Collected:	The questionnaire asks if substance abuse treatment is offered in different languages at the facility. The question includes the following choices: American/Alaska Native languages (Hopi, Lakota, Navajo, Yupik, Other American/Alaska Native language), and Other languages (Arabic, Chinese, Creole, French, German, Hmong, Korean, Polish, Portuguese, Russian, Spanish, Vietnamese, and Other language).

CDC SURVEILLANCE DATA SYSTEMS

(in alphabetical order)

Adult Blood-Lead Epidemiology and Surveillance Program (ABLES)

Agency:	National Institute for Occupational Safety and Health
Purpose of Collection:	To reduce the number of adults with

blood lead levels of 25 mcg/dl or greater.

Source of Data: Laboratory blood level reports

Universe of Data Collection: A sample of adults in 35 states

Racial and Ethnic Data Collected: Not collected

SEP Data Collected: Not collected

Language and Acculturation
Data Collected: Not collected

Adult Spectrum of Disease (ASD) (HIV Patients)

Agency: National Center for HIV, STD, and TB Prevention

Purpose of Collection: To enumerate and characterize persons with HIV.

Source of Data: Patient medical records

Universe of Data Collection: All persons with HIV infection who access selected hospitals, outpatient facilities, and HIV treatment facilities in the 10 selected project areas.

Racial and Ethnic Data Collected: American Indian/Alaska Native; Asian or Pacific Islander; Black, Not of Hispanic Origin; White, Not of Hispanic Origin; Hispanic

SEP Data Collected: Not collected

Language and Acculturation
Data Collected: Not collected

Childhood Blood Lead Surveillance

Agency: National Center for Environmental Health

Purpose of Collection: To build state capacity to conduct surveillance; to establish a national surveillance system based on state systems; and to use data to direct

	prevention activities at the local, state, and national levels.
Source of Data:	Laboratory slips
Universe of Data Collection:	Children aged 16 years or younger who participate in surveys or whose cases are reported to the state or local health department
Racial and Ethnic Data Collected:	Native American/Alaska Native; Asian/Pacific Islander (Asian Indian, Chinese, Filipino, Hawaiian, Korean, Vietnamese, Japanese, Samoan, Hmong, Guamanian, Other, Unknown); Black; White; Multiracial; Other; Unknown. Ethnic categories are: Hispanic; Non-Hispanic; Unknown. The quality of these data varies among states. The primary source of the data is laboratory slips, which often contain incomplete data on race and ethnicity.
SEP Data Collected:	Not collected
Language and Acculturation Data Collected:	Not collected

Congenital Syphilis Cases Investigation and Report (CSCIR)

Agency:	National Center for HIV, STD, and TB Prevention
Purpose of Collection:	To provide detailed data on congenital syphilis cases.
Source of Data:	Patient records
Universe of Data Collection:	All mothers and infants that are reported to CDC as having syphilis by STD control programs and health departments in all states, DC, selected cities, and U.S. dependencies and possessions
Racial and Ethnic Data Collected:	American Indian/Alaska Native; Asian

or Pacific Islander; Black; White; Other

Ethnic categories are: Hispanic; Not of Hispanic Origin

SEP Data Collected: Not collected

Language and Acculturation Data: Not collected

Creutzfeldt-Jakob Disease Surveillance System (CJD)

Agency: National Center for Infectious Diseases

Purpose of Collection: To collect brain autopsy material from persons with hemophilia and other bleeding disorders who had received care in treatment centers anywhere in the United States. This material is then examined to determine whether the disease can be transmitted through blood or blood products.

Source of Data: Patient medical records

Universe of Data Collection: Persons with hemophilia and other bleeding disorders who died and who had received care in treatment centers anywhere in the United States.

Racial and Ethnic Data Collected: White (non-Hispanic), White (Hispanic), Black (non-Hispanic), Asian/Pacific Islander, American Indian/Alaska Native, Other.

SEP Data Collected: Not collected

Language and Acculturation Data Collected: Not collected

Diphtheria Antitoxin (DAT)

Agency: National Immunization Program

Purpose of Collection: To characterize cases of diphtheria and thereby identify risk factors for

the disease and areas at risk for
outbreaks.

Source of Data: Diphtheria antitoxin is only available
                                     through CDC. As such, it must be
                                     released on a case-by-case basis. CDC
                                     collects information on suspected
                                     diphtheria cases in the process of drug
                                     release and in follow-up via telephone
                                     discussions with treating physicians.

Universe of Data Collection: All persons in the United States. who
                                     are or are suspected to be infected
                                     with diphtheria

Racial and Ethnic Data Collected: Uses OMB standards

SEP Data Collected: Not collected

*Language and Acculturation
   Data Collected:*   Not collected

Firearm Injury Surveillance Study

Agency: National Center for Injury Prevention
                                     and Control

Purpose of Collection: To understand the magnitude and
                                     characteristics of nonfatal firearm-
                                     related injuries in the United States.

Source of Data: Medical records

Universe of Data Collection: Persons with firearm injuries who
                                     receive treatment at United States
                                     hospitals that provide emergency
                                     services.

Racial and Ethnic Data Collected: American Indian/Alaska Native; Asian
                                     or Pacific Islander; Black; White;
                                     Other. Ethnic categories are: His-
                                     panic, Not of Hispanic Origin. Data
                                     on race and ethnicity are obtained as
                                     specified on hospital emergency
                                     department records. If a person is
                                     reported as Hispanic, race is usually
                                     recorded as "other."

SEP Data Collected: Not collected

Language and Acculturation
  Data Collected: Not collected

Foodborne Disease Active Surveillance Network (FoodNet)

Agency: National Center for Infectious Dis-
                                         eases

Purpose of Collection: To help public health officials better
                                         understand the epidemiology of
                                         foodborne disease in the United
                                         States.

Source of Data: Surveys, patient records

Universe of Data Collection: Persons in the United States who have
                                         contracted a foodborne disease

Racial and Ethnic Data Collected: Available, but categories are unspeci-
                                         fied.

SEP Data Collected: Not collected

Language and Acculturation
  Data Collected: Not collected

Gonococcal Isolate Surveillance Project: Demographic/Clinical Data and Antimicrobial Susceptibility Testing (GISP)

Agency: National Center for HIV, STD, and
                                         TB Prevention

Purpose of Collection: To monitor trends in antimicrobial
                                         susceptibilities of strains of *N.
                                         gonorrhoeae* in the United States in
                                         order to establish a rational basis for
                                         the selection of gonococcal therapies.

Source of Data: Gonococcal isolates, patient records

Universe of Data Collection: *N. gonorrhoeae* isolates are collected
                                         from the first 25 men attending STD
                                         clinics each month in 26 cities in the
                                         United States.

Racial and Ethnic Data Collected: American Indian/Alaska Native; Asian or Pacific Islander; Black; White; Other. Ethnic categories are: Hispanic; Not of Hispanic Origin. Data on race may not be collected at each site.

SEP Data Collected: Not collected

Language and Acculturation Data Collected: Not collected

Haemophilus Influenzae Surveillance System (HI)

Agency: Epidemiology Program Office, National Immunization Program, and National Center for Infectious Diseases

Purpose of Collection: To compile information on all U.S. Haemophilus influenzae invasive disease cases reported to CDC via the National Electronic Telecommunications System for Surveillance (NETSS) since 1991, or via active surveillance in several locales since 1989.

Source of Data: Patient medical records

Universe of Data Collection: All U.S. Haemophilus influenzae invasive disease cases reported by health departments and hospitals

Racial and Ethnic Data Collected: The OMB Statistical Directive 15 two-variable standard for reporting is used. Racial categories are: White; Black; Asian/Pacific Islander; American Indian/Alaska Native; Other. Ethnic categories are: Hispanic/Latino origin; not of Hispanic/Latino origin.

SEP Data Collected: Not collected

Language and Acculturation Data Collected: Not collected

Hemophilia Surveillance System (HSS)

Agency:	National Center for Infectious Diseases
Purpose of Collection:	To identify all persons with hemophilia in six states (CO, GA, LA, MA, NY, OK) and to characterize the population according to demographic and clinical features.
Source of Data:	Medical records
Universe of Data Collection:	All persons with hemophilia in six states
Racial and Ethnic Data Collected:	Racial and/or ethnic data are available using the following categories: White (non-Hispanic), White (Hispanic), Black (non-Hispanic), Asian/Pacific Islander, American Indian/Alaska Native, Other.
SEP Data Collected:	Education
Language and Acculturation Data Collected:	Not collected

HIV/AIDS Reporting System (HARS)

Agency:	National Center for HIV, STD, and TB Prevention
Purpose of Collection:	To monitor the total number of reported cases of HIV/AIDS from public, private, and government reporting facilities.
Source of Data:	Patient records
Universe of Data Collection:	All reported AIDS cases in the 50 states, territories, and possessions and HIV cases in states that require reporting of persons with HIV (not AIDS)
Racial and Ethnic Data Collected:	American Indian/Alaska Native; Asian or Pacific Islander; Black, Not of

Hispanic Origin; White, Not of Hispanic Origin; Hispanic

SEP Data Collected: Not collected

Language and Acculturation
   Data Collected: Not collected

HIV Seroprevalence Studies

Agency: National Center for HIV, STD, and TB Prevention

Purpose of Collection: To monitor HIV seroprevalence among different groups of people, such as intravenous drug users, childbearing women, and people who visit STD clinics.

Source of Data: Patient records

Universe of Data Collection: People entering drug treatment programs at 53 centers in 22 U.S. cities

Racial and Ethnic Data Collected: American Indian/Alaska Native; Asian or Pacific Islander; Black; White; Other. Ethnic categories are: Hispanic (Mexican-American, Puerto Rican, Cuban); Not of Hispanic Origin

SEP Data Collected: Not collected

Language and Acculturation
   Data Collected: Not collected

Laboratory Surveillance of Cryptosporidium and Malaria via Public Health Laboratory Information Systems (PHLIS)

Agency: National Center for Infectious Diseases

Purpose of Collection: To identify persons with Cryptosporidium and malaria.

Source of Data: Patient record

Universe of Data Collection: All persons in the United States with Cryptosporidium or malaria

Racial and Ethnic Data Collected: American Indian or Alaska Native; Asian or Pacific Islander; Black; White; Unknown. Ethnic categories are: Hispanic; Non-Hispanic; Unknown.

SEP Data Collected: Not collected

Language and Acculturation
   Data Collected: Not collected

National Bacterial Meningitis Reporting System (NBMRS)

Agency: National Center for Infectious Diseases

Purpose of Collection: To identify persons in the United States with bacterial meningitis.

Source of Data: Patient record

Universe of Data Collection: All persons in the United States with bacterial meningitis

Racial and Ethnic Data Collected: Available, but the categories are unspecified.

SEP Data Collected: Not collected

Language and Acculturation
   Data Collected: Not collected

National Campylobacter Surveillance System

Agency: National Center for Infectious Diseases

Purpose of Collection: To learn more about how Campylobacter causes disease and is spread.

Source of Data: Patient record

Universe of Data Collection: All persons in the United States with campylobacterlosis, which is caused by Campylobacter isolates in the body.

Racial and Ethnic Data Collected: Not collected

SEP Data Collected: Not collected

Language and Acculturation
  Data Collected: Not collected

National Malaria Surveillance System (NMSS)

Agency: National Center for Infectious Dis-
                                         eases

Purpose of Collection: To identify individuals with malaria
                                         in the United States.

Source of Data: Medical records

Universe of Data Collection: All persons with malaria in the United
                                         States

Racial and Ethnic Data Collected: Available, but categories are not
                                         listed.

SEP Data Collected: Occupation

Language and Acculturation
  Data Collected: Not collected

National Nosocomial Infections Surveillance System (NNIS)

Agency: National Center for Infectious Dis-
                                         eases

Purpose of Collection: To describe the epidemiology of
                                         nosocomial infections, to describe
                                         antimicrobial resistance trends, and to
                                         produce nosocomial infection rates to
                                         use for comparison purposes.

Source of Data: Laboratory charts

Universe of Data Collection: All persons in the United States with
                                         nosocomial infections

Racial and Ethnic Data Collected: Not collected

SEP Data Collected: Not collected

Language and Acculturation
  Data Collected: Not collected

National Program of Cancer Registries (NPCR)

Agency: National Center for Chronic Disease Prevention and Health Promotion

Purpose of Collection: To rapidly establish and standardize the reporting of cancer among the states in order to provide: (1) timely feedback for evaluating progress toward achieving cancer-control objectives that include the "Healthy People 2010" objectives; (2) data to monitor the incidence and mortality trends in patterns by age, ethnicity, and geographic regions within the state, between states, and between regions; (3) guidance for health resource allocation; (4) data to evaluate state cancer-control activities; and (5) information to improve planning for future health care needs.

Source of Data: Patient record

Universe of Data Collection: All persons in the United States with cancer

Racial and Ethnic Data Collected: White; Black; American Indian/Aleutian or Eskimo; Chinese; Japanese; Filipino; Hawaiian; Korean; Asian Indian/Pakistani; Vietnamese; Laotian; Hmong; Kampuchean; Thai; Micronesian, NOS; Chamorran; Guamanian, NOS; Polynesian, NOS; Tahitian; Samoan; Tongan; Melanesian, NOS; Fiji Islander; New Guinean; Other Asian (including Asian, NOS and Oriental, NOS); Pacific Islander, NOS; Other; Unknown.

Ethnic categories are: Spanish/Hispanic Origin; Non-Spanish/Non-Hispanic; Mexican (includes Chicano); Puerto Rican; Cuban; South or Central American (except

Brazilian); Other Spanish (includes European); Spanish, NOS/Hispanic, NOS/Latino, NOS (i.e., there is evidence other than surname or maiden name that the person is Hispanic, but he/she cannot be assigned to any of the preceding categories); Spanish (surname only) (only evidence of person's Hispanic origin is surname or maiden name); Unknown Whether Spanish or Not

SEP Data Collected: Not collected

Language and Acculturation
  Data Collected: Not collected

National Salmonella Surveillance System (NSSS)

Agency: National Center for Infectious Diseases

Purpose of Collection: To identify the incidence of salmonella infection in the United States

Source of Data: Electronic data reports

Universe of Data Collection: All persons in the United States infected with salmonella

Racial and Ethnic Data Collected: Not available

SEP Data Collected: Not collected

Language and Acculturation
  Data Collected: Not collected

National Surveillance for
Domestic Arboviral Encephalitis Cases in Humans

Agency: National Center for Infectious Diseases

Purpose of Collection: To monitor arboviral encephalitis cases and reporting practices in the United States.

Source of Data: Patient records

Universe of Data Collection: All persons in the United States in-
                                        fected with arboviral encephalitis

Racial and Ethnic Data Collected: Native American/Alaska Native;
                                        Asian/Pacific Islander; Afro American;
                                        White; Other; Unknown. Ethnic
                                        categories are: Hispanic; Not His-
                                        panic; Unknown

SEP Data Collected: Not collected

Language and Acculturation
  Data Collected: Not collected

National Surveillance of Nonfatal Occupational Injury

Agency: National Institute for Occupational
                                        Safety and Health

Purpose of Collection: To obtain a national overview of
                                        nonfatal occupational injuries, to
                                        study specific types of injuries, to
                                        study specific worker populations,
                                        and to meet a variety of other occupa-
                                        tional injury research needs.

Source of Data: Patient records

Universe of Data Collection: All civilians in the United States who
                                        suffer nonfatal work-related injuries.

Racial and Ethnic Data Collected: White; Black; Other (additional
                                        information may be included in a
                                        free-text field); Not stated. Ethnicity
                                        may be included in a free-text field
                                        for Race = Other. Not all of the
                                        participating hospitals collect racial
                                        and ethnic data.

SEP Data Collected: Information about the person's indus-
                                        try

Language and Acculturation
  Data Collected: Not collected

Pertussis

Agency: National Immunization Program

Purpose of Collection: To provide details of each pertussis
                                       case reported by each state in the
                                       United States

Source of Data: Patient records/interviews

Universe of Data Collection: All persons in the United States with
                                       pertussis

Racial and Ethnic Data Collected: Uses OMB standards

SEP Data Collected: Not collected

Language and Acculturation
*   Data Collected:*   Not collected

Plague

Agency: National Center for Infectious Dis-
                                       eases

Purpose of Collection: To identify presumptive and con-
                                       firmed cases of plague.

Source of Data: Patient records

Universe of Data Collection: All persons presumed or confirmed to
                                       have plague

Racial and Ethnic Data Collected: Racial and/or ethnic data are avail-
                                       able. There have only been about 400
                                       plague cases over the past 40 years.
                                       The racial and ethnic categories have
                                       been used for the past four decades.
                                       Changes from the original categories
                                       are in parentheses below. No effort
                                       has been made to update them to fit
                                       current styles. The categories are:
                                       White (or Caucasian, Non-Hispanic);
                                       Hispanic (or Caucasian, Hispanic);
                                       Indian (now Native American, listed
                                       by tribe whenever possible, e.g., AIN
                                       = American Indian Navajo, American
                                       Indian Pueblo, etc.); Oriental (now

would be referred to as Asian—only one such case); Pacific Islander (only one such person—from the Philippines). There are no black cases on the list. If such a case did occur, however, it would be recorded as black for race/ethnicity.

SEP Data Collected: Not collected

Language and Acculturation Data Collected: Not collected

Sentinel Counties Surveillance for Acute Study of Viral Hepatitis

Agency: National Center for Infectious Diseases

Purpose of Collection: To monitor incidence trends and transmission patterns for hepatitis A, hepatitis B, hepatitis C, and other emerging viral hepatitis infections. In addition, the system is used to monitor the effectiveness of prevention and control programs for these diseases.

Source of Data: Patient records

Universe of Data Collection: All persons in the United States infected with viral hepatitis

Racial and Ethnic Data Collected: White (not of Hispanic origin); Black (not of Hispanic origin); American Indian or Alaska Native; Asian or Pacific Islander; Hispanic; Unknown

SEP Data Collected: Education and income

Language and Acculturation Data Collected: Not collected

Sentinel-Site Laboratory-Based Surveillance for Cyclospora

Agency: National Center for Infectious Diseases

Purpose of Collection: To identify Cyclospora in eight states
                                   and the District of Columbia.

Source of Data: Laboratory samples

Universe of Data Collection: All cases of Cyclospora in the United
                                   States

Racial and Ethnic Data Collected: Not collected

SEP Data Collected: Not collected

Language and Acculturation
   Data Collected: Not collected

State-Based Emergency Department Injury Surveillance

Agency: National Center for Injury Prevention
                                   and Control

Purpose of Collection: To capture the statewide incidence of
                                   emergency department-related inju-
                                   ries.

Source of Data: Medical records

Universe of Data Collection: All persons in the United States who
                                   suffer an injury that causes them to
                                   visit an emergency department

Racial and Ethnic Data Collected: Not collected

SEP Data Collected: Not collected

Language and Acculturation
   Data Collected: Not collected

STD Surveillance Systems

Agency: National Center for HIV, STD, and
                                   TB Prevention

Purpose of Collection: To collect data on cases of syphilis
                                   (all stages), congenital syphilis (before
                                   1995), gonorrhea, chancroid, and
                                   chlamydia.

Source of Data: Patient medical records

Universe of Data Collection: People diagnosed with any of the

above diseases who visit private
doctors or nurses, hospitals, and
public health clinics, and data from
public health laboratories and state/
local health departments

Racial and Ethnic Data Collected: Not collected

SEP Data Collected: Not collected

Language and Acculturation
* Data Collected:*   Not collected

Streptococcus Pneumoniae and Haemophilus Influenzae

Agency: National Center for Infectious Dis-
                             eases

Purpose of Collection: Patterns and rates of disease cases are
                             used to monitor and modify preven-
                             tion strategies to examine risk factors
                             for disease, and to monitor trends in
                             the development of antimicrobial
                             resistance of isolates.

Source of Data: Patient records

Universe of Data Collection: All persons in Alaska with Streptococ-
                             cus Pneumoniae and Haemophilus
                             Influenzae.

Racial and Ethnic Data Collected: Collected, but categories are not
                             specified.

SEP Data Collected: Not collected

Language and Acculturation
* Data Collected:*   Not collected

Surveillance for Giardia

Agency: National Center for Infectious Dis-
                             eases

Purpose of Collection: To identify cases of giardia within the
                             United States.

Source of Data: Patient records

Universe of Data Collection:	All persons in the United States with giardia.
Racial and Ethnic Data Collected:	American Indian or Alaska Native; Asian or Pacific Islander; Black; White; Unknown. Ethnic categories are: Hispanic; Non-Hispanic; Unknown.
SEP Data Collected:	Not collected
Language and Acculturation Data Collected:	Not collected

Surveillance for Pneumocystis Carinii Pneumonia Treatment Failures

Agency:	National Center for Infectious Diseases
Purpose of Collection:	To study treatment outcome among appropriately treated laboratory-confirmed Pneumocystis Carinii Pneumonia in HIV-positive individuals.
Source of Data:	Patient records
Universe of Data Collection:	All persons in the United States with HIV who are treated for Pneumacystis Carinii Pneumonia
Racial and Ethnic Data Collected:	White, Non-Hispanic; Black, Non-Hispanic; Hispanic/Latino; American Indian/Alaska Native; Asian/Pacific Islander; Other; Unknown
SEP Data Collected:	Not collected
Language and Acculturation Data Collected:	Not collected

Surveillance for Trichinosis

Agency:	National Center for Infectious Diseases
Purpose of Collection:	To monitor trends in trichinosis.

Source of Data: Patient records

Universe of Data Collection: All persons in the United States with
                                   trichinosis

Racial and Ethnic Data Collected: American Indian or Alaska Native;
                                   Asian or Pacific Islander; Black;
                                   White; Unknown. Ethnic categories
                                   are: Hispanic; Non-Hispanic; Un-
                                   known.

SEP Data Collected: Not collected

Language and Acculturation
   Data Collected: Not collected

Surveillance for Tuberculosis Infection in Health Care Workers (StaffTRAK-TB)

Agency: National Center for HIV, STD, and
                                   TB Prevention

Purpose of Collection: To test health care workers for TB
                                   infection routinely as per CDC infec-
                                   tion control guidelines.

Source of Data: Direct interview/self-report

Universe of Data Collection: All health care workers from partici-
                                   pating hospitals

Racial and Ethnic Data Collected: Asian, Black, Hispanic, American
                                   Indian/Alaska Native, White, Other
                                   races

SEP Data Collected: Occupation

Language and Acculturation
   Data Collected: Country of birth

Tetanus

Agency: National Immunization Program

Purpose of Collection: To provide data describing each
                                   tetanus case with respect to the
                                   wound, type of wound, and where it
                                   occurred.

Source of Data:	Patient records
Universe of Data Collection:	All persons in the United States with tetanus
Racial and Ethnic Data Collected:	Uses OMB standards
SEP Data Collected:	Occupation
Language and Acculturation Data Collected:	Not collected

Traumatic Brain Injury Surveillance System (TBISS)

Agency:	National Center for Injury Prevention and Control
Purpose of Collection:	To understand the magnitude and characteristics of hospitalized and fatal traumatic brain injuries in the United States; to collect program wide information that can be used to help design targeted prevention programs addressing specific causes of traumatic brain injuries and populations at higher risk; to improve injured persons' access to health care; and to improve other services they need after injury.
Source of Data:	Patient records
Universe of Data Collection:	All persons in the United States who suffer a traumatic brain injury
Racial and Ethnic Data Collected:	American Indian/Alaska Native; Asian or Pacific Islander; Black; White; Other. Categories of ethnicity are Hispanic and Not of Hispanic Origin. Since information on nonfatal cases is obtained from hospital discharge data systems, data on race/ethnicity are not available from states.
SEP Data Collected:	Not collected
Language and Acculturation Data Collected:	Not collected

Tuberculosis Information Management System (TIMS)

Agency: National Center for HIV, STD, and
                                     TB Prevention

Purpose of Collection: To assist health departments and
                                     other facilities to manage TB patients,
                                     to conduct TB surveillance activities,
                                     and to manage TB programs overall.

Source of Data: Direct interview and patient records

Universe of Data Collection: All persons with TB who visit private
                                     doctors or nurses, hospitals, and
                                     public health clinics, and data from
                                     public health laboratories and state/
                                     local health departments

Racial and Ethnic Data Collected: Uses OMB standards

SEP Data Collected: Not collected

Language and Acculturation
  Data Collected: Not collected

Unexplained Deaths and Serious Illnesses Surveillance

Agency: National Center for Infectious Dis-
                                     eases

Purpose of Collection: To identify and collect clinical speci-
                                     mens from people who have suffered
                                     unexplained deaths or serious ill-
                                     nesses.

Source of Data: Medical records

Universe of Data Collection: All previously healthy persons in the
                                     United States, aged 1-49 years, who
                                     have died for no obvious reason or
                                     who have experienced an unexplained
                                     serious illness

Racial and Ethnic Data Collected: Available, but categories are not
                                     specified

SEP Data Collected: Occupation

Language and Acculturation
  Data Collected: Not collected

Universal Data Collection and Serum Specimen Collection System (UDC)

Agency:	National Center for Infectious Diseases
Purpose of Collection:	To collect prospective clinical data on persons with hemophilia and other bleeding disorders in the United States who receive at least some of their care from federally sponsored hemophilia treatment centers.
Source of Data:	Medical records
Universe of Data Collection:	All persons in the United States with hemophilia and other bleeding disorders who receive at least some of their care from federally sponsored hemophilia treatment centers
Racial and Ethnic Data Collected:	Available, but no standardized categories are used
SEP Data Collected:	Education
Language and Acculturation Data Collected:	Not collected

United States Influenza Sentinel Physician Surveillance Network

Agency:	National Center for Infectious Diseases
Purpose of Collection:	To identify how many people in the United States have influenza or an influenza-like illness.
Source of Data:	Physician survey
Universe of Data Collection:	All persons who visit a doctor in the United States and exhibit symptoms of influenza
Racial and Ethnic Data Collected:	Not collected
SEP Data Collected:	Not collected
Language and Acculturation Data Collected:	Not collected

Vaccine Adverse Event Reporting System (VAERS)

Agency: National Immunization Program

Purpose of Collection: To compile and monitor national estimates of adverse events by vaccine antigen and vaccine lot.

Source of Data: Patient records

Universe of Data Collection: All persons in the United States who experience an adverse event due to vaccination

Racial and Ethnic Data Collected: Not collected

SEP Data Collected: Not collected

Language and Acculturation Data Collected: Not collected

Viral Hepatitis Surveillance Program (VHSP)

Agency: National Center for Infectious Diseases

Purpose of Collection: To compile clinical, serologic, and epidemiologic data on cases of hepatitis A, B, and C and to monitor the effectiveness of prevention and control programs for these diseases.

Source of Data: Patient records

Universe of Data Collection: All persons in the United States who have hepatitis A, B, or C

Racial and Ethnic Data Collected: American Indian or Alaska Native; Asian or Pacific Islander; Black; White; Unknown. Ethnic categories are: Hispanic; non-Hispanic; Unknown.

SEP Data Collected: Not collected

Language and Acculturation Data Collected: Place of birth, either United States. or Other (specify), and place of birth of the person's mother

Waterborne Diseases Outbreak Surveillance System (WBDOSS)

Agency:	National Center for Infectious Diseases
Purpose of Collection:	To collect and summarize data on waterborne disease outbreaks and identify trends in outbreaks summarized by the type of contaminant and by the type of water that was contaminated.
Source of Data:	Survey and patient records
Universe of Data Collection:	All persons in the United States infected with a waterborne disease
Racial and Ethnic Data Collected:	Not collected
SEP Data Collected:	Not collected
Language and Acculturation Data Collected:	Not collected

HUMAN SERVICE PROGRAMS IN DHHS

Child Care and Development Block Grant

Agency:	Administration for Children and Families
Purpose:	To make grants to states and tribes to assist low-income families with child care and to: (1) allow each state maximum flexibility in developing child care programs and policies that best suit the needs of children and parents within state; (2) promote parental choice to empower working parents to make their own decisions on the child care that best suits their family's needs; (3) encourage states to provide consumer education information to help parents make informed choices about child care; (4) assist states to provide child care to parents trying to achieve independence from

public assistance; and (5) assist states in implementing the health, safety, licensing, and registration standards established in state regulations.

Population Covered:

Children under the age of 13 (or up to age 19 if physically or mentally incapable of self-care or under court supervision) who reside with a family whose income does not exceed 85 percent of the state median income for a family of the same size, and who reside with a parent (or parents) who is working or attending job training or an educational program; or who are in need of, or are receiving protective services.

Racial and Ethnic Data Collected: Uses OMB standards

SEP Data Collected:

Total monthly family income and sources of family income are collected.

Language and Acculturation
   Data Collected:

Not collected

Child Support Enforcement Program (CSE)

Agency:

Administration for Children and Families

Purpose:

This program is a federal/state/local partnership that works to locate noncustodial parents, establish paternity when necessary, establish orders for support, and collect child support payments for families.

Population Covered:

Services are available to a parent with custody of a child who has a parent living outside of the home. Services are available automatically for families receiving assistance under the Temporary Assistance for Needy Families (TANF) program. Current

child support collected reimburses the
state and federal governments for
TANF payments made to the family.
Those not receiving public assistance
can apply for child support services.
Child support payments that are
collected on behalf of non-TANF
families are sent to the family.

Racial and Ethnic Data Collected: Not collected

SEP Data Collected: Not collected

*Language and Acculturation
  Data Collected:* Not collected

Adoption and Foster Care Analysis and Reporting System

Agency: Administration for Children and
Families

Purpose: Collects information on all children in
foster care for whom the state child
welfare agency has responsibility as
well as children who are adopted
under the auspices of the state's
public child welfare agency.

Population Covered: All children in foster care

Racial and Ethnic Data Collected: Uses OMB standards (can choose
more than one race)

SEP Data Collected: Not collected

*Language and Acculturation
  Data Collected:* Not collected

Head Start

Agency: Head Start Bureau, Administration
for Children and Families

Purpose: To increase the school readiness of
young children in low-income fami-
lies.

Population Covered:	Children from low-income families, from birth to age 5
Racial and Ethnic Data Collected:	Uses OMB standards
SEP Data Collected:	Household income
Language and Acculturation Data:	Language spoken at home and a rating of English proficiency are gathered.

Maternal and Child Health Block Grant (MCH)

Agency:	Maternal and Child Health Bureau, Health Resources and Services Administration
Purpose:	To promote and improve the health of all United States mothers and children by supporting states' efforts to extend and improve health and welfare services for mothers and children.
Population Covered:	All mothers and children in the United States
Racial and Ethnic Data Collected:	Uses OMB standards (can choose more than one race)
SEP Data Collected:	Income
Language and Acculturation Data Collected:	Not collected

The National Child Abuse and Neglect Data System

Agency:	Administration for Children and Families
Purpose:	To provide state and national data on child abuse and neglect for program planning, program review, and policy development.
Population Covered:	All children in the United States
Racial and Ethnic Data Collected:	Uses OMB standards

SEP Data Collected: Not collected

Language and Acculturation
  Data Collected: Not collected

National Survey of Child and Adolescent Well-Being

Agency: Administration for Children and
                                         Families

Purpose: To examine child and family well-
                                         being outcomes in detail and seek to
                                         relate those outcomes to their experi-
                                         ence with the child welfare system
                                         and to family characteristics, commu-
                                         nity environment, and other factors.
                                         The study describes the child welfare
                                         system and the experiences of children
                                         and families who come in contact
                                         with the system.

Population Covered: Children who are at risk of abuse or
                                         neglect or are in the child welfare
                                         system

Racial and Ethnic Data Collected: Uses OMB standards

SEP Data Collected: Information about financial resources
                                         available to the child's family is
                                         collected.

Language and Acculturation
  Data Collected: Not collected

Social Services Block Grant (SSBG)

Agency: Administration for Children and
                                         Families

Purpose: To fund states for the provision of
                                         social services directed toward achiev-
                                         ing economic self-support or self-
                                         sufficiency, preventing or remedying
                                         neglect, abuse, or the exploitation of
                                         children and adults, preventing or
                                         reducing inappropriate institutional-

ization, and securing referral for institutional care, where appropriate.

Population Covered:	All people in the United States
Racial and Ethnic Data Collected:	Not collected
SEP Data Collected:	Not collected .
Language and Acculturation Data Collected:	Not collected

Temporary Assistance to Needy Families (TANF)

Agency:	Office of Family Assistance, Administration for Children and Families
Purpose:	To provide assistance and work opportunities to needy families by granting states with federal funds and wide flexibility to develop and implement their own welfare programs.
Population Covered:	Low-income people in the United States
Racial and Ethnic Data Collected:	Uses OMB standards
SEP Data Collected:	Collects employment, income, and cash resources information
Language and Acculturation Data Collected:	Not collected

Uses OMB standards: American Indian and Alaska Native; Asian; Black or African American; Native Hawaiian and Other Pacific Islander; White. Respondents shall be allowed the option of selecting one or more racial categories. There are two categories for data on ethnicity: Hispanic or Latino and Not Hispanic or Latino.

Appendix B

Workshop on Improving Racial and Ethnic Data in Health

AGENDA

December 12-13, 2002

Thursday, December 12, 2002

10:00 **Welcome and Introductions**

       Edward Perrin, University of Washington (Panel Chair)
       Andrew White, Committee on National Statistics
       James Scanlon, Office of Science and Data Policy, Office of the
            Assistant Secretary for Planning and Evaluation, Department
            of Health and Human Services

10:15 **The Importance of Collecting Data on Race, Ethnicity, and
       Socioeconomic Status**

       Session Chair: **David Williams,** University of Michigan (Panel
          Member)
       Authors: **Nicole Lurie** and **Allen Fremont,** RAND

10:45 Discussants:
       Olivia Carter-Pokras, University of Maryland
       Gem Daus, Asian and Pacific Islander American Health Forum

11:15 Open Discussion

11:30   **Legal Aspects of Race and Ethnicity Data Collection (Including Privacy Issues)**
         Session Chair: **William Kalsbeek,** University of North Carolina, Chapel Hill (Panel Member)

11:30   Legal Framework: **Mara Youdelman,** National Health Law Program

12:00   Privacy Issues: **Moya Gray,** Director, Hawaii Office of Information Practices

12:30   Open Discussion

1:45   **Race and Ethnicity Data Collection by Private Organizations**
         Session Chair: **L. Carl Volpe,** WellPoint Health Networks Inc. (Panel Member)

1:45   **General Overview—Collection of Data on Race/Ethnicity of Patients by Hospitals, Medical Groups, and Health Plans**
         Author: **David Nerenz,** Michigan State University

2:15   *The Collection of Race and Ethnicity Data by Health Plans*
         Author: **Carmella Bocchino,** American Association of Health Plans

3:00   Discussants:
         Eileen Peterson, United Health Group
         Romana Hasnain-Wynia, Health Research and Education Trust
         Andrew Webber, National Committee for Quality Assurance

4:00   Open Discussion
         Moderator: **Edward Perrin,** University of Washington (Panel Chair)

Friday, December 13

8:30   **Race and Ethnicity Data Collection by States**
         Session Chair: **Denise Love,** National Association of Health Data Organizations (Panel Member)
         Authors: **Sara Singer** and **Jeff Geppert,** Stanford University; **Jay Buechner,** Rhode Island; **Wu Xu,** Utah; **Walter Suarez,** Minnesota; and **Lorin Ranbom,** Ohio

9:00   Discussants:
         Dan Friedman, Massachusetts Department of Public Health
         Carla Edwards, Florida Agency for Health Care Administration
         David Solet, Public Health, Seattle and King County

9:30   Open Discussion

10:00   **Attendee Comments on Race and Ethnicity Data Collection**
         Session Chair: **Anthony D'Angelo**, Computercraft (Panel
             Member)
         Workshop attendees were invited to sign up to give brief
             comments on the collection of racial and ethnic data

11:15   **Perspectives of Data Users**
         Session Chair: **Jonathan Skinner**, Dartmouth College (Panel
             Member)
         Panelists:
         Matthew Snipp, Stanford University
         Rhonda BeLue, Metro Nashville Public Health Department
         Marian Gornick, Consultant, Health Services Research

WORKSHOP PARTICIPANTS LIST

Presenters

Edward Perrin (*Panel Chair*), University of Washington
Rhonda BeLue, Metro Nashville Public Health Department
Carmella Bocchino, American Association of Health Plans
Jay Buechner, Rhode Island Department of Health
Olivia Carter-Pokras, University of Maryland, Baltimore Campus
Anthony D'Angelo, Computercraft
Gem Daus, Asian and Pacific Islander American Health Forum
Carla Denise Edwards, Florida Agency for Health Care Administration
Allen Fremont, RAND
Dan Friedman, Massachusetts Department of Public Health
Jeffrey Geppert, Stanford University
Marian Gornick, Health Services Research
Moya Gray, Hawaii Office of Information Practices
Romana Hasnain-Wynia, Health Research and Education Trust
William Kalsbeek, University of North Carolina, Chapel Hill
Denise Love, National Association of Health Data Organizations
Nicole Lurie, RAND
David Nerenz, Michigan State University
Eileen Peterson, United Health Group
Lorin Ranbom, Ohio Department of Health
James Scanlon, Office of Science and Data Policy, Department of Health
    and Human Services
Sara Singer, Stanford University
Jonathan Skinner, Department of Economics, Dartmouth College

Matthew Snipp, Stanford University
David Solet, Public Health, Seattle and King County
Walter Suarez, Minnesota Department of Health
L. Carl Volpe, Strategic Health Partnership, WellPoint Health Networks Inc.
Andrew Webber, National Committee for Quality Assurance
Andrew White, Committee on National Statistics, National Academies
David Williams, Department of Sociology, University of Michigan
Wu Xu, Utah, Department of Health
Mara Youdelman, National Health Law Program

Invited Guests

Steven Auerbach, Health Resources and Services Administration
Roman Baez, Dental School Multicultural Programs
John Burton, Council on African American Affairs, Inc.
Audrey Burwell, Department of Health and Human Services
Shari Campbell, Bureau of Primary Health Care
Marcia Changkit, National Cancer Institute
Kenneth Chu, Center for Reduce Cancer Health Disparities
Beverly Colman-Miller,
Pam Curry, DHHS Office of Counter Terrorism and Pediatric Drug Development
Charles Daly, Bureau of Primary Health Care
Shelia Davis, Summit Health Institute for Research and Education
Shirley Delaleu, Council on African American Affairs, Inc.
Elaine Elinksy, Elinsky Consulting
Roland Garcia, National Cancer Institute
Kenneth Gerlach, Cancer Surveillance Branch
Linda Greenberg, Centers for Medicare and Medicaid Services
Robert Hahn, Centers for Disease Control and Prevention
Jan Heffernan, National Institutes of Health
Katherine Hollinger, Office of Women's Health, FDA
Diane Hopkins, U.S. Department of Health and Human Services
Jin In, U.S. Department of Health and Human Services
Debbie Jackson, National Center for Health Statistics, CDC
K.A. Jagannathan, DHHS Administration for Children and Families
Cheedy Jaja, Indiana University Center for Bioethics
Yvonne Johns, DHHS Office of Minority Health
Steve Jones II, United States Congress
Evelyn Kappeler, DHHS Office of Population Affairs
Onelio Lopez, U.S. Department of Health and Human Services

Jennifer Madans, Centers for Disease Control
Diane Makuc, National Center for Health Statistics
Jennifer Malat, University of Cincinnati
Mildred Martinez, Kaiser Permanente
Donielle Newell, Council on African American Affairs, Inc.
Ly Nguyen, Morgan State University
Edna Paisano, DHHS Office of Public Health
JoAnn Pappalardo, DHHS Office of Public Health
Sharyn Parks, National Center for Injury Prevention and Control
William Rodriguez, DHHS Office of Counter Terrorism and Pediatric
    Drug Development
Beatrice Rouse, Substance Abuse Mental Health Services Administration
Theodore Small Jr., Council on African American Affairs, Inc.
Bryon Sogie-Thomas, National Medical Association
Edward Sondik, Centers for Disease Control
Irene Tsai, Kaiser Permanente
Luz Vega, District of Columbia Department of Health
Irene Zimmerman, Washington Association of Professional
    Anthropologists

NRC Staff

Michele Ver Ploeg, Committee on National Statistics
Jamie Casey, Committee on National Statistics
Tanya Lee, Committee on National Statistics
Andrew White, Committee on National Statistics

Appendix C

Recommendations on the Use of Socioeconomic Position Indicators to Better Understand Racial Inequalities in Health

Patricia O'Campo and Jessica Burke *

INTRODUCTION

The purpose of this paper is to explore what socioeconomic position measures can and should be collected along with racial and ethnic data to measure and better understand disparities in health.

Substantial research documents racial and ethnic disparities in health status and health care access. Racial disparities in disease incidence have been demonstrated for many health outcomes including cardiovascular disease, HIV/AIDS, diabetes, and infant morality (U.S. Department of Health and Human Services, 2002). For example, the national infant mortality rate (IMR) for African Americans is 2.5 times that of whites, with an IMR as high as 13.8 deaths per 1,000 live births for African Americans (Hoyert et al., 2000). The mortality rates for 8 of the 10 leading causes of death are much higher among African American/black populations as compared to whites (National Center for Health Statistics, 2000). The Institute of Medicine's (IOM) recent report *Unequal Treatment: Confronting Racial and Ethnic Disparities in Health Care* is one of several documents that chronicle research highlighting racial inequalities in health (Institute of Medicine, 2003). A comprehensive discussion of racial inequalities in health is beyond the scope of this paper, but additional information can be found

*Patricia O'Campo, Ph.D., and Jessica Burke, Ph.D., are professor and research associate, respectively, at the Johns Hopkins Bloomberg School of Public Health in Baltimore, MD.

in the IOM report and other sources (Williams, 1999; Collins, Hall, and Neuhaus, 1999; LaVeist, 2002; Krieger et al., 1993).

Evidence also supports a strong relationship between socioeconomic position (SEP) and both health status and health care access (Lynch and Kaplan, 2000; Krieger, Williams, and Moss, 1997). Explanations for the relationship between SEP and health tend to focus on how differences in material circumstances and in health-related knowledge and behavior are associated with differences in health outcomes (Townsend and Davidson, 1992; Lynch and Kaplan, 2000). Literature examines the relationships between pathways of SEP and health outcomes by suggesting that single indicators of SEP (i.e., education, occupation, or income and wealth) are interconnected (Krieger, Williams, and Moss, 1997). Specifically, education is an important determinant of future employment opportunities, and occupational status is related to earning potential, income, and future wealth. Income, in turn, can directly influence health by enabling the procurement of material circumstances associated with positive health outcomes. For example, individuals with higher incomes are more likely than their poorer counterparts to have the financial means to afford to live in a clean and safe environment and to afford health care insurance.

But the amount of money associated with an individual is only part of the picture. Knowledge of health-related issues and of where and how to seek health care, both typically associated with increased educational status, are also important factors contributing to health outcomes.

While much discussion has taken place regarding the endogeneity of health and socioeconomic position, evidence supporting a link between poor health and decreased SEP is limited and inconsistent across life stages. Instead, research has found stronger support for the pathway of lower socioeconomic position leading to worse health outcomes (Manor, Matthews, and Power, in press; McDonough and Amick, 2001).

It is also well recognized that race and socioeconomic position are highly interrelated and that both of these factors are strongly associated with health status (Adler et al., 1994; Krieger, Williams, and Moss, 1997; Williams, 1999). However, epidemiologists and other public health researchers in general fail to adequately account for the role of SEP factors in racial inequalities research. Socioeconomic position is often not taken into account at all. National Vital Statistics Reports produced by the Centers for Disease Control and Prevention are just one example of multiple official health reports that present data only on the racial distribution of health outcomes and fail to simultaneously address SEP issues (e.g., Anderson, 2002). While other epidemiologic studies find that racial disparities are attenuated once adjustments have been made to account for socioeconomic position, often such approaches are inadequate. Specifically, use of single indicators of socioeconomic position such as education or category of income does not fully account for economic differences between the groups (Kaufman, Cooper, and McGee, 1997). For example, Schoendorf and colleagues (1992), in a study among college-educated parents, showed that racial disparities in infant mortality rates persisted despite accounting for the SEP indicator of education. Findings from this study show an IMR of 10.2 per 1,000 live births for black infants and a rate of 5.4 per 1,000 live births for white infants. The likelihood of death for a black infant was thus 1.82 times that of a white infant even after controlling for age and parity (Schoendorf et al., 1992). In the article's conclusion, the authors acknowledge that the persistent gap in IMR may be attributable to economic and social differences between the black and white samples that were not addressed.

There are several problems inherent in the use of single indicators or crude measures to control or adjust for the impact of socioeconomic position that make it less than the ideal approach when studying racial inequalities in health (Kaufman, Cooper, and McGee, 1997). The concept of SEP is complex, and one-dimensional measures (e.g., education) do not fully capture it. In addition, the experiences associated with standard SEP indicators (income, education, and occupation) are not the same among different racial groups. For example, whites have approximately twice the income, four times the net financial assets, and a staggering nine times the net worth of blacks (Oliver and Shapiro, 1997). In addition, black men are more likely than white men to be employed in jobs that expose them to hazards and carcinogens (Robinson, 1984). In the few studies that have adjusted for a fuller range of socioeconomic position indicators, a gap in health between racial and ethnic groups often remains. For example, data from the National Longitudinal Mortality Survey (Muntaner, Sorlie, and O'Campo, 2001) showed unadjusted odds ratios for cardiovascular disease mortality for black men and women compared to whites to be 1.5 for men and 2.0, for women. Adjustment for several socioeconomic indicators, such as education, income, and occupational status, resulted in a reduction of those odds ratios to 1.3 and 1.8. Still, an unexplained gap in cardiovascular disease mortality for blacks compared to whites remained after accounting for numerous indicators of social class.

Adjustment using a single or crude indicator of socioeconomic position results in problems of residual confounding for economic position when comparing health status or health care utilization between racial or ethnic groups. Use of multiple indicators will minimize the degree to which residual confounding is a problem. However, as noted earlier, some reports of health status or health care utilization do not adjust for any SEP indicators (e.g., NCHS reports of U.S. live births or infant deaths). The reporting of health status by only racial or ethnic group gives the erroneous impression that the within-group heterogeneity may be less than the between-group heterogeneity with respect to multiple factors including socioeconomic position. Under such circumstances, meaningful inferences about racial inequalities in health cannot be made. Therefore, when possible, adjustment with any SEP indicator is preferable to no adjustment at all.

It is apparent that greater efforts need to be made to identify and utilize sound measures of socioeconomic position in racial disparities research. While several solid reviews of SEP indicators have been published in recent years (Liberatos, Link, and Kelsey, 1988; Krieger, Williams, and Moss, 1997; Lynch and Kaplan, 2000; Duncan et al., 2002; Oaks and Rossi, 2003; Berkman and Macintyre, 1997), they offer little insight into which measures are best to use when studying racial disparities in health. For example, the recent review by Oaks and Rossi (2003) provides a history of SEP research, highlights the paucity of research on the measurement of socioeconomic position, and suggests that composite measures may offer the most insight into the complex relationship between SEP and health. However, they do not ultimately offer recommendations about which measures to use in general, or specifically in health disparities research. Although Nazroo's (2003) discussion of social and economic inequalities as the fundamental causes of racial disparities in health draws attention to our limited understanding of these causal relationships and points out the limited availability of SEP data in large routine administrative data sets, he too fails to offer insight into which existing measures of SEP might best be incorporated in studies of racial health disparities. General SEP research has found that economic measures (i.e., income and wealth) outperform measures of education and occupation (Lynch and Kaplan, 2000; Duncan et al., 2002). Yet, given the racial differences in wealth described above, additional work is necessary before concluding that these findings are also applicable to health disparities research.

In order to adequately address how socioeconomic position is related to racial inequalities in health status, researchers must take significant strides toward identifying and incorporating appropriate measures in their investigations (Kaufman, Cooper, and McGee, 1997). In the following sections, we first summarize our findings from a review of SEP measures included in nationally available administrative and survey sources. We then provide an overview of various SEP indicators. And finally we offer recommendations about which measures to utilize in health disparities research.

Before proceeding, a brief discussion of the term *race* as used in this paper is necessary. Within public health there is much disagreement about the term. Often, public health researchers mistakenly base their notions of race on the idea that the human species can be separated into distinct human races identifiable through differences in physical characteristics. Setting the standards for the nation, the U.S. Office of Management and Budget (OMB) recommended using the categories of American Indian or

Alaska Native, Asian, Black or African American, Native Hawaiian or Other Pacific Islander, and White. The OMB recognizes that these categories represent a "sociopolitical construct" and "are not anthropologically or scientifically based." We agree. In this paper, our interpretation and use of the term *race* refers to socially determined, not biological, characteristics.

SOCIOECONOMIC POSITION INDICATORS FOUND IN ADMINISTRATIVE AND SURVEY SOURCES

We reviewed common health-related administrative and survey sources to determine the availability of socioeconomic position indicators. For each source, we noted the methods of identifying information, the inclusion or absence of geographic information, the frequency of data collection, the SEP indicators included, and the coding of each SEP measure. A summary of our findings can be found in Table C-1. All of the reviewed health-related administrative and survey sources, with the exception of the Medicare Program Data, include a measure of at least one SEP indicator. Consistent with prior reviews (Lynch and Kaplan, 2000; Duncan et al., 2002; Oaks and Rossi, 2003), we found that educational status was the most frequently cited measure of socioeconomic position. Multiple surveillance systems and surveys also include measures of both income and education. For example, the Medicare Current Beneficiary Survey includes indicators of income and education. While not a direct measure of income, source of payment (e.g., Medicare, Blue Cross/Blue Shield, self-pay, etc.) is also collected by several national health surveys such as the National Hospital Discharge Survey. This source-of-payment variable can sometimes serve as a proxy for income status. Only three of the sources reviewed address issues of wealth, and none utilize individual-level indices of socioeconomic position or indicators of social class.

Among the administrative and survey sources presented in Table C-1, the use of multiple SEP indicators varies dramatically. Administrative sources, in general, tend to rely on single or limited indicators of socioeconomic position. For example, the National Vital Statistics System and the Consumer Assessment of Health Plans both rely on education as the sole indicator of socioeconomic position. National surveys, on the other hand, tend to include more than one measure of socioeconomic position. For example, the National Survey of Early Childhood Health includes questions regarding maternal education and employment, and family income. Two prime examples of more comprehensive approaches to determining socioeconomic position can be seen in the Medical Expenditure Panel Survey (MEPS) and the National Health and Nutrition Examination Survey (NHANES). In addition to detailed questions about employment status (e.g., name of employer, type of employer, job title, job duties), the MEPS

TABLE C-1 Measures of Socioeconomic Position from Survey and Administrative Data for Analysis of Racial Inequalities in Health

INDICATORS OF INDIVIDUAL OR HOUSEHOLD SOCIAL STRATIFICATION

Wealth	Short set of questions: ownership of housing and value of home if sold on the market at the time of the survey, and value of any owned vehicles. If net financial assets are sought, then inquire further about the amount of the mortgage principal that remains to be paid on the home or loan for the car.
	Detailed questions: information on housing ownership as above as well as the value of any other real estate property owned; vehicle ownership and value; business ownership and value; investment income and value; value of savings and other banking accounts. If net financial assets are sought, must also obtain information on debt for all assets owned.
Income	Short set of questions: annual, monthly, weekly or hourly wages of the respondent or household expressed in continuous dollars or categories.
	Short set of questions: annual, monthly, weekly or hourly wages of the respondent's household adjusted for family size expressed in continuous dollars or categories.
	Detailed set of questions: income from all sources including wages and salary; investment income; government program participation; rental property income; business income (Duncan et al., 2002).
Poverty	Using information on family or household income adjusted for family or household size, can compare this value to the poverty threshold for that year to obtain the percent poverty expressed as a continuous or categorical variable.
Education	Number of years of formal schooling completed by the respondent or client.
	Educational credentialing (i.e., degrees such as high school diploma, college diploma, etc.).
Individual level indices	Examples of such indices that combine information on education, occupation and sometimes income include: Hollingshead Index of Social Position, which combines information on an individual's educational attainment and occupational ranking (Liberatos, Link, and Kelsey, 1988) and Duncan's Socioeconomic Index of occupational prestige (Duncan, 1961).
Government program participation	Examples of program participation includes: unemployment, Temporary Assistance for Needy Families, Food Stamps, General Assistance, WIC, Section 8 housing and government assistance for transportation or utilities or child care.
	Health insurance status and/or method of payment for health care Insurance types include: private insurance, Medicare, Medicaid/SCHIP, self-pay Sources of payment include: worker's compensation, Medicaid, Medicare, private or commercial insurance, HMO/PPO, self-pay.

continued

TABLE C-1 Continued

INDICATORS OF INDIVIDUAL SOCIAL CLASS

Wright's categorization	Combined information on ownership assets (employer large firm, employer small firm, self-employed, or nonowner), organizational assets (manager, supervisor, nonmanagement), and skills/credentials assets (experts, marginal, uncredentialed) (see Appendix D; Wright, 1997).

AREA LEVEL MEASURES OF SOCIOECONOMIC POSITION

Wealth	Proportion home ownership or proportion with annual incomes greater than $50,000.
Poverty	Concentrated poverty where greater than 40 percent of the households live at or below poverty (Wilson, 1996) or other appropriate thresholds.
Occupational class	Proportion of employed persons who fall into specific occupational categories such as professional and managerial positions or proportion falling into working class occupations (e.g., administrative support; sales; private household; other service; precision production, craft, repair; machine operators, assemblers, inspectors; transportation and material moving; handlers, equipment cleaners, laborers).
Education	Proportion of adults with greater than a high school degree.
Indices	Townsend index: a combination of unemployment among those 16-64; proportion of households with no car; proportion of households who do not own their dwellings; overcrowding as measured by more than one person per room. Other indices might include the Socioeconomic Status Index, which combines information on education and income of an area, and Stockwell, which differentially weights information on occupation, education, income, housing values, and overcrowding (see Liberatos, Link, and Kelsey, 1988; or Krieger, 1999; Krieger, Williams, and Moss, 1997, for overviews of various composite measures).

also asks about sources of income and the value of assets that the respondent may hold. NHANES respondents are questioned about their education, employment (including type of business/industry and type of work), income, type of health insurance, and homeownership.

For many of the administrative and survey sources it is possible to link individual-level records to area-level SEP indicators (Krieger, Williams, and Moss, 1997). For example, the National Vital Statistics System includes mother's residential address with the birth record file. This address information can be mapped to specific residential area designations (e.g., census tract) and then assigned area-level indicators of socioeconomic position using available administrative data such as those from the census. While a few of the databases included in this review do not contain respondent street address information (e.g., National Hospital Discharge Survey), a residential Zip Code is provided for each respondent and can be used in a similar fashion to tie individual records to area-level information.

REVIEW OF INDICATORS OF SOCIOECONOMIC POSITION

We describe here the specific indicators of socioeconomic position and how they might be used to increase our understanding of racial inequalities in health. Strengths, weaknesses, and comparisons of the indicators of socioeconomic position are described below. Details about measuring SEP indicators are contained in Table C-1.

Wealth

For reasons noted earlier, wealth information is probably the most critical data to collect if one is attempting to account for economic differences between racial and ethnic groups. Numerous reports and studies have documented that the greatest differences between racial and ethnic groups are wealth differences and not educational or income differences. Blacks and Hispanics have one-tenth the household net worth of whites (U.S. Bureau of the Census, 1996). Educational attainment does not account for this difference as even among college-educated persons, blacks have 23 percent the household net worth of whites (Oliver and Shapiro, 1995).

Questions about wealth capture information, at a single point in time, about household financial assets (e.g., housing ownership, ownership of rental properties and businesses, investment income, vehicles, etc.). Net financial assets account for the household debt on these assets. Duncan and Petersen (2001, Appendix C) outline recommendations for short and long data collection instruments for wealth and discuss important aspects of ensuring high-quality complete data on wealth and income that should be considered by those collecting this information on surveys. For example, if a respondent answers "don't know" to a question about housing value, this answer can be followed by other questions to attempt to get the respondent to give an approximation (e.g., "Would the house sell for more or less than $50,000 if it were sold today?").

Information about wealth is not commonly requested in studies nor collected in administrative data sources because of perceptions and experiences indicating that it is both difficult to collect and subject to nonreporting. However, given that most U.S. residents have cars and homes as their sole assets, a short set of questions might ask only about these two items. (Many U.S. households also have retirement income, but for those under 65 that income is not available to use.) Duncan and Petersen recommend collecting information on housing in a short set of questions. The utility of this short set will depend on the population being studied. For example, a great majority of higher-income individuals and families are likely to own homes or cars and such information may not differentiate disadvantaged higher-income groups as well as it might in lower-income populations. More detailed questions about wealth tap information on several factors as noted in Table C-1. This level of detail is possible only in surveys and not normally available from administrative sources. However, among those administrative sources that do collect information on income, another question on housing ownership and value might be considered. When possible, to promote a better understanding of the interaction of class and racial inequalities in health, information on wealth should be collected in future surveys and in administrative sources of data. The collection of wealth data, even through a short set of wealth questions, would greatly enhance our ability to understand racial and ethnic inequalities and health.

Income and Poverty

Income usually refers to wages and salary received either by a person (i.e., individual income) or by all members of a family or household (i.e., household income). Less frequently, a more comprehensive assessment of income from all sources (e.g., salary, government programs, business income, income from rental property, etc.) is collected for individuals or households (Krieger, Williams, and Moss, 1997; Duncan and Petersen, 2001).

While many researchers report that income data are difficult to obtain and suffer from a high proportion of missing information, Duncan and Petersen (2001) discuss ways to promote the reporting of this information (see above discussion regarding wealth). For example, for surveys, Duncan and Petersen (2001) do not recommend having the respondent choose from categories of salary and wages. Rather, a direct question on the dollar amount is recommended. Then, "unfolding scales" and prompts that follow answers of "don't know" can be used to reduce missing information (Duncan and Petersen, 2001). These options are not always possible to use with administrative sources of data.

A simple set of indicators can reflect total income for either an individual or a household. Household is probably preferable to individual income as it is a reflection of the current economic resources available to the household or family. Information on individual income, on the other hand, because it varies widely within families, may not accurately reflect the current resources available to that individual (e.g., the income of a respondent who works part-time might appear artificially low if other adults in the household also work. Adjusting household income for the number of members being supported also increases the accuracy of the measure of current economic resources available to that individual. The economic resources available to each family member in a family of two that earns $40,000 annually differ from those available to members of a family of four that earns the same amount.

Poverty is an alternative indicator to income. Poverty can be determined using information on salary and wages and household size. The advantage of using poverty indicators is that, when expressed in relation to the official poverty line, they provide a measure of whether the household has the minimum resources for basic necessities. Although the official poverty measure underestimates the amount a family needs to meet basic necessities, it is standardized and can be compared across studies. While not commonly used, income can alternatively be expressed in terms of other economic metrics such as a living wage, which may better reflect how close or far families are from being able to afford basic necessities. It should also be noted that indicators such as income and poverty show greater fluctuations from year to year than an indicator such as wealth (Stevens, 1999). If a stable indicator of economic standing is desired, wealth may be preferable to income or poverty status.

Education

Education is the most commonly measured indicator of socioeconomic position. Some have argued that it is the best indicator because of its stability over the lifecourse as well as the ease and accuracy for which these data can be collected (Winkleby et al., 1992). However, the fact that education is usually stable once a person has passed early adulthood may be one of its major weaknesses in that it does not capture the volatility that is present in the lives of many, especially the poor (e.g., Stevens, 1999). Moreover, it does not directly reflect the economic resources available to a person or household as income and wealth indicators do. Finally, use of education as the sole indicator of socioeconomic position is likely to yield significant residual confounding given the variability in income and wealth seen between racial and ethnic groups for the same levels of educational attainment (e.g., Oliver and Shapiro, 1995).

Despite these limitations, education is often the only SEP indicator available (e.g., in administrative data sources) and collection of education data is relatively easy and subject to little misclassification. These advantages should be kept in mind when options are very limited in terms of survey or interview length.

Education data can be collected as a continuous variable (i.e., number of years of schooling completed), which can be categorized (see Table C-1) or, in addition, in terms of credentials obtained (i.e., high school diploma, college degree, master's degree, etc.). The latter is important because credentials are often the means to advancement in the workplace. Yet for purposes of adjustment between racial and ethnic groups, the availability of information on both continuous education and credentials would be ideal.

Social Class

Stratification indicators, the most commonly used measures of socioeconomic position in the United Stated, are based on Weberian notions of how power, privilege, resources, and prestige give persons differential access to life opportunities (Krieger, Williams, Moss, 1997; Muntaner et al., 2000; Lynch and Kaplan, 2000). While indicators of social class are rare in the U.S. literature, they yield important information about *how* health inequalities are created by the opposing interests of those in different classes (Krieger, Williams, and Moss, 1997; Muntaner, Eaton, and Diala, 2000; Lynch and Kaplan, 2000). Class struggle—"the struggle between such collectively organized actors over class interests"—is critical to understanding how inequalities between classes are generated (Wright, 1997). There is ample evidence of class struggle all around us in policies like the scaling back of social programs such as welfare, health care coverage, or unemployment benefits, or an increase in jobs without benefits, or a low, poverty-level, minimum wage for the nation. Because of this explicit link to mechanisms by which class processes can create inequalities, these social class indicators yield information that is different from the usual stratification measures (Muntaner, Eaton, and Diala, 2000).

The set of questions developed by Erik Olin Wright (1997) has been the most commonly implemented set of social class measures in the health literature. If our goal is to better understand the reasons for racial and ethnic inequalities in health we may want to consider including and using indicators of social class more frequently in future studies and analytic efforts. Questions on social class indicators could easily be included in surveys, but it is unlikely that this full set of data would be feasibly obtained from administrative sources.

Individual-Level Indices

Researchers have combined indicators on individual education, occupation, and income to create indices of socioeconomic position. While these indicators were developed and have been used primarily in sociology, some public health researchers have used them as well. Examples of the latter include the Hollingshead Index of Social Position (Hollingshead, 1975), Duncan's Socioeconomic Index (Duncan, 1961), and, for occupational specific indicators, the Nam-Powers Occupational Status Score (Nam and Powers, 1983). Because these indices, especially those that are occupationally based, were created decades ago and combine information from different indicators of socioeconomic position, their utility has been questioned (Liberatos, Link, and Kelsey, 1988; Krieger, Williams, and Moss, 1997). Indices such as those used in Nam-Powers are based on occupational prestige ratings that were devised several decades ago and that therefore may no longer be relevant to the current job market.

The collection of these data is more burdensome than the collection of the separate components alone (e.g., income, education). This should be kept in mind when time or interview length are limited.

Combining different indicators of socioeconomic position both precludes the identification of which factors may matter most and limits the interpretation of the overall index in terms of inequalities between racial and ethnic groups. Therefore, the use of multiple, separate indicators of SEP is recommended over the use of these indices.

Government Program Participation

Government program participation information, typically collected in administrative sources, is useful for the study of lower-income populations as, for program eligibility, there are often income cutoffs (e.g., Temporary Assistance for Needy Families, Food Stamps, Medicaid), not though for all assistance programs (e.g., Medicare, unemployment benefits). Thus, in the absence of specific information on economic resources, this information might be useful in determining which individuals or families are of low income. The information cannot, however, be used to directly estimate income as eligibility requirements differ by program and often across states.

Health Insurance Type or Source of Payment for Medical Care

Health insurance and payment information, like government program participation, can be used to categorize individuals or households as lower or higher income. Some specific types of insurance, primarily Medicaid, are based on low-income eligibility. Therefore, this information might serve as a proxy to indicate income level, with private insurance indicating higher incomes. However, given the ever-changing insurance situation in the United States, these categories do not always clearly correlate with income levels; for example, middle- or high-income individuals or families may be uninsured or self-pay for medical care. But in the absence of other direct income information, these indicators can be used to suggest relative income levels of a sample.

Area-Level Indicators of Social Position

Recently there has been increasing interest in using area-level indicators of social position (Krieger, Williams, and Moss, 1997; Lynch and Kaplan, 2000). Area-level indicators might be used in place of or in addition to individual data on social position. The advantages and disadvantages of this method have been discussed in depth in the literature (Kaufman, Cooper, and McGee, 1997; Krieger, Williams, and Moss, 1997; Geronimus, Bound, and Neidert, 1996; Mustard et al., 1999). There are several reasons for briefly mentioning this methodology here. First and foremost, given that many administrative sources lack sufficient characterization of social position (e.g., they may have only the educational status of the individual) but may have information on the home address that can be used to link to area-level data, this method may be a means by which individual social position data may be augmented with area-level social position information. Such a link may be particularly important when comparing racial or ethnic differences, as ample data have shown vast differences in the socioeconomic conditions of neighborhoods of different racial and ethnic groups (Jargowsky, 1997; Massey and Denton, 1993). Another reason is that if the goal for one type of comparison between racial and ethnic groups is to adjust as fully as possible for socioeconomic position, then more than one indicator, and indicators at more than one level, may be desirable for this purpose. Kaufman, Cooper, and McGee (1997) warn of the need to use these measures as continuous (e.g., continuous levels of income rather than as a fixed category of, say, greater than $20,000) in order to retain as much information as possible for this very purpose. Table C-1 thus includes several economic-, occupation-, and education-based indicators that can be combined to represent neighborhood socioeconomic position. To create these indicators, the person in the data survey or administrative base must simply provide information on residential address, which can then be used to link to census (or other administrative) sources of information to estimate the indicators.

USING SEP INDICATORS TO UNDERSTAND RACIAL INEQUALITIES IN HEALTH STATUS AND USE OF HEALTH CARE SERVICES

To fully understand racial inequalities in health, we must begin to do a better job of untangling SEP from racial group membership. Thus far, few studies or reports have adequately accounted for SEP in a way that makes racial or ethnic groups comparable. Currently, there is a wide range of practices that account for SEP when comparing the health of racial groups. Given that the indicators described in the previous section and in Table C-1 cannot act as proxies for one another, careful thought must go into selection of the single or multiple indicators to be used. For example, one cannot collect information on income and assume that it will serve as a good proxy for wealth or that education can serve as a proxy for income. The correlations of these measures have been examined and found to be moderate at best (Parker, Schoendorf, and Kiely, 1994; Braveman et al., 2001). Moreover, the correlation of socioeconomic position variables differs by race or ethnicity (Braveman et al., 2001). One must carefully identify the reason for the collection and measurement of particular SEP indicators, as described earlier in the paper, and collect information, when possible, on all indicators needed. Here we offer brief guidance as to which indicators might be given priority for collection in research surveys or for administrative data sets. We base these recommendations on the review of administrative data sources and the descriptions of SEP indicators and their advantages and disadvantages described in the previous sections.

We begin by discussing the socioeconomic indicators for existing administrative data sources. It was encouraging to see that most of the administrative data sources we reviewed contain indicators of SEP that can be used when examining racial and ethnic inequalities in health. While education is the most common SEP indicator collected in these administrative data sets, some sources also contained information on critical economic indicators such as income and even wealth. For the data sets that contain only education, additional collection of a more direct indicator of economic standing (e.g., income or wealth) would facilitate an even greater understanding of the nature of the racial inequalities we now often observe.

Administrative data sets with no information on SEP are the real challenge. It is critical that efforts be undertaken to collect SEP information. Where possible, the collection of more than one indicator should be considered. If the ability to do so is limited, then simple questions on education, participation in government programs, or the short set of questions on income might be considered. Another option is to use participant residential address to link census-based measures of SEP for purposes of analysis and reporting; this approach avoids the need to collect additional information in instances where it is just not possible.

In enrollment or service delivery encounters where there are often constraints on collecting more than one indicator of SEP, the collection of any SEP indicator is preferable to no such information. While indicators of economic standing (e.g., wealth and income) are desirable if time for data collection permits, information on client education or government program participation may be easier to collect, is less sensitive, and may require less time. If address information is available, it can be linked to residential area socioeconomic data.

Given that research surveys often have the most flexibility for including questions on SEP we suggest that, when possible, multiple indicators be collected, with data on wealth receiving the highest priority. This approach of collecting multiple indicators allows for residual confounding to be addressed in the analyses of survey data on racial and ethnic differences in health. Moreover, if analyses are conducted on single groups, the availability of data on multiple indicators facilitates a more complete understanding of socioeconomic variation within a particular racial or ethnic group that may contribute to adverse health outcomes. Because surveys that include multiple indicators of socioeconomic standing will contribute the greatest insights into the magnitude and reasons for racial inequalities in health, placing greater emphasis on appropriate measurement in surveys of SEP is critical to future research efforts on racial and ethnic inequalities in health.

Finally, while the indicators discussed in this paper emphasize measures of SEP at a single point in time (i.e., at the time the survey is completed) it is recommended that, when possible, SEP information be collected for multiple points in time (Lynch and Kaplan, 2000). Childhood SEP—for example, the experience of long-term or multiple spells of poverty during childhood—can have consequences for later health status (Marmot and Wilkinson, 1999). Also, SEP during early to middle adulthood can have effects on health status for the elderly independent of their SEP after retirement. Therefore, where possible, most likely for research surveys, information on SEP for multiple time points across the lifespan should be collected.

ACKNOWLEDGMENT

The authors would like to acknowledge Adam Allston's contribution to this work in compiling the table of socioeconomic status indicators available in national surveillance systems and health surveys contained in Table C-1.

REFERENCES

Anderson, R.N.
    2002   Deaths: Leading causes for 2000. *National Vital Statistics Reports* 50(16).
Berkman L., and S. Macintyre
    1997   The measurement of social class in health studies: Old measures and new formulations. Pp. 51-64 in M. Kogevinas, N. Pearce, and P. Boffetta, eds. *Social Inequalities and Cancer.* Lyon, France: International Agency for Research on Cancer.
Braveman P., C. Cubbin, K. Marchi S., Egerter, and G. Chavez
    2001   Measuring socioeconomic status/position in studies of racial/ethnic disparities: Maternal and infant health. *Public Health Reports* 116(5):449-463.
Collins, K.S., A. Hall, and T. Neuhaus
    1999   *U.S. Minority Health: A Chartbook.* New York: The Commonwealth Fund.

Duncan G., and E. Petersen
    2001   The long and short of asking questions about income, wealth and labor supply. *Social Science Research* 30:243-263
Duncan, G.J., M.C. Daly, P. McDonough, and D. Williams
    2002   Optimal indicators of socioeconomic status for health research. *American Journal of Public Health* 92(6):1151-1157.
Duncan, O.
    1961   A socioeconomic index for all occupations. In W. Reiss, ed., *Occupations and Social Status*. New York: Free Press.
Geronimus, A., J. Bound, and L. Neidert
    1996   On the validity of using census geocode characteristics to proxy individual socio-economic characteristics. *Journal of the American Statistical Association* 91:517-529.
Hollingshead, A.
    1975   *Four Factor Index of Social Status*. New Haven, CT: Department of Sociology, Yale University.
Hoyert, D.L., M.A. Freedman, D.M. Strobino, and B. Guyer
    2000   Annual summary of vital statistics. *Pediatrics* 108(6):1241-55
Institute of Medicine
    2003   *Unequal Treatment: Confronting Racial and Ethnic Disparities in Health Care.* B.D. Smedley, A.Y. Stith, and A.R. Nelson, eds. Committee on Understanding and Eliminating Racial and Ethnic Disparities in Health Care. Board on Health Sciences Policy. Washington, DC: The National Academies Press.
Jargowsky, P.A.
    1997   *Poverty and Place: Ghettos, Barrios and the American City.* New York: Russell Sage Foundation.
Kaufman, J.S., R.S. Cooper, and D.L. McGee
    1997   Socioeconomic status and health in Blacks and Whites: The problem of residual confounding and the resiliency of race. *Epidemiology* 8:621-628.
Krieger, N.
    1999   Embodying inequality: A review of concepts, measures, and methods for studying health consequences of discrimination. *International Journal of Health Services* 29(2):295-352.
Krieger, N., D.L. Rowley, A.A. Herman, B. Avery, and M.T. Phillips
    1993   Racism, sexism, and social class: Implications for studies of health, disease, and well-being. *American Journal of Preventive Medicine* 9:82-122.
Krieger, N., D.R. Williams, and E. Moss
    1997   Measuring social class in U.S. public health research: Concepts, methodologies and guidelines. *Annual Review of Public Health* 18:341-378.
LaVeist, T., ed.
    2002   *Race, Ethnicity and Health: A Public Health Reader.* San Francisco, CA: Jossey-Bass.
Liberatos, P., B. Link, and J. Kelsey
    1988   The measurement of social class in epidemiology. *Epidemiologic Reviews* 10:87-121.
Lynch, J., and G. Kaplan
    2000   Socioeconomic position. In L. Berkman and I. Kawachi, eds., *Social Epidemiology.* New York: Oxford University Press.
Manor, O., S. Matthews, and C. Power
    in   Health selection: The role of inter- and intra-generational mobility on social
    press   inequalities in health. *Social Science and Medicine.*

Marmot, M., and R. Wilkinson, eds.
   1999   *Social Determinants of Health.* New York: Oxford University Press.
Massey, D., and N. Denton
   1993   *American Apartheid: Segregation and the Making of the Underclass.* Cambridge,
          MA: Harvard University Press.
McDonough, P., and B.C. Amick III
   2001   The social context of health selection: A longitudinal study of health and employ-
          ment. *Social Science and Medicine* 53:135-145.
Muntaner, C., W. Eaton, and C. Diala
   2000   Social inequalities in mental health: A review of concepts and underlying assump-
          tions. *Health* 4:89-113.
Muntaner C., P. Sorlie, and P. O'Campo
   2001   Finance, government and production work and cardiovascular mortality among
          men and women: Findings from the National Longitudinal Mortality Study. *Annals
          of Epidemiology* 11(3):194-201.
Mustard, C., S. Derksen, J. Berthelot, and M. Wolfson
   1999   Assessing ecologic proxies for household income: A comparison of household and
          neighborhood level income measures in the study of population and health status.
          Health and Place 5:157-171.
Nam, C., and M. Powers
   1983   *The Socioeconomic Approach to Status Measurement with a Guide to Occupa-
          tional and Socioeconomic Status Scores.* Houston: Cap and Gown Press.
National Center for Health Statistics
   2000   *Health, United States, 1999, with Socioeconomic Status and Health Chartbook.*
          Hyattsville, MD: National Center for Health Statistics.
Nazroo, J.Y.
   2003   The structuring of ethnic inequalities in health: Economic position, racial discrimi-
          nation and racism. *American Journal of Public Health* 93(2):277-284.
Oaks, J.M., and P.H. Rossi
   2003   The measurement of SES in health research: Current practice and steps toward a
          new approach. *Social Science and Medicine* 56(4):769-784.
Oliver, M., and T. Shapiro
   1995   *Black Wealth / White Wealth: A New Perspective in Racial Inequality.* New York:
          Routledge.
Parker J., K. Schoendorf, and J. Kiely
   1994   Associations between measures of socioeconomic status and low birth weight, small
          for gestational age, and premature delivery. *United States Annals of Epidemiology*
          4:271-278
Robinson, J.
   1984   Racial inequality and the probability of occupation-related injury or illness. *Milbank
          Quarterly* 62:567-590.
Schoendorf, K.C., C.J.R. Hogue, J.C. Kleinman, and D. Rowley
   1992   Mortality among infants of black as compared with white college-educated par-
          ents. *New England Journal of Medicine* 326:1522-1526.
Stevens A.
   1999   Climbing out of poverty, falling back in: Measuring the persistence of poverty over
          multiple spells. *Journal of Human Resources* 34(3):557-588.
   1995   *Asset Ownership of Households, 1993.* Washington, DC: U.S. Government Print-
          ing Office.

Townsend, P., and Davidson, N.
  1992   The Black report. In P. Townsend, M. Whitehead, and N. Davidson, eds., *Inequality in Health: The Black Report and Health Divide*. London, England: Penguin.
U.S. Bureau of the Census
  1996   *Money Income in the United States: 1995*. Washington, DC: U.S. Bureau of the Census.
U.S. Department of Health and Human Services
  2002   *Healthy People 2010*. Available: http://web.health.gov.
Williams, D.R.
  1999   Race, socioeconomic status, and health: The added effect of racism and discrimination. *Annals of New York Academy of Sciences* 896:173-188.
Wilson, W.J.
  1996   *When Work Disappears: The World of the New Urban Poor*. New York: Alfred Knopf.
Winkleby, M., D. Jatulis, E. Frank, and S. Fortmann
  1992   Socioeconomic status and health: How education, income, and occupation contribute to risk factors for cardiovascular disease. *American Journal of Public Health* 82(6):816-820.
Wright E.
  1997   Class Counts. Cambridge, England: Cambridge University Press.

Appendix D

The Role of Racial and Ethnic Data Collection in Eliminating Disparities in Health Care

Allen Fremont and Nicole Lurie*

Racial and ethnic disparities in health have been extensively documented. While the causes are both numerous and diverse, disparities in health care have been shown to play a substantial role. A recent Institute of Medicine report (IOM, 2003) exhaustively catalogued disparities in care and concluded that important differences were present even among groups that were similarly insured. Many observers now conclude that eliminating racial and ethnic disparities in health and health care is a central issue in overall efforts to improve quality (IOM, 2001; Bierman et al., 2002). Consistent with this view, *Healthy People 2010* specified elimination of such disparities as one of its two overarching goals (U.S. Department of Health and Human Services, 2000b).

Making progress toward the goal of eliminating disparities will require widespread, reliable, and consistent data about the racial and ethnic characteristics of the U.S. population. This information is needed to identify the nature and extent of disparities, to target quality improvement efforts, and to monitor progress. Tracking the racial and ethnic composition and changing health care needs of different populations is vital if our health care system, which includes both public health and the delivery of personal health care services, is to fulfill its essential functions. Measurement, reporting, and benchmarking are critical to improving care.

---

*Allen Fremont, M.D., Ph.D., and Nicole Lurie, M.D., M.S.P.H, are a medical sociologist and primary care physician, and senior natural scientist and Paul O'Neill Alcoa Professor, respectively, at the RAND Corporation, in Arlington, VA.

Despite widespread public perception that the federal government and the private sector collect vast amounts of data, the availability of racial and ethnic data in the health care system itself is quite limited. A variety of government sources include data on race and ethnicity, but the utility of these data is constrained by ongoing problems with reliability, completeness, and lack of comparability across data sources. With only a few exceptions, private insurers and health plans do not maintain data on the race or ethnicity of their enrollees.

In this paper, we provide a framework for describing the role of racial and ethnic data in supporting essential functions of the health system. We first illustrate the value of racial ethnic data collection by describing ways such information can be used to reduce disparities, particularly with respect to the quality of care. We describe how data on primary language and socioeconomic status can complement racial and ethnic information. We then assess current sources of racial and ethnic information and the challenges inherent in collecting it. We conclude with a series of recommendations for enhancing the availability and use of data in the public and private sectors.

Throughout our discussion, we emphasize the federal role in data collection. However, since only about half of the minority population in the United States receives care in a public-sector system (e.g., Medicare, Medicaid, Department of Veterans Affairs, Department of Defense), the importance of private-sector and other government efforts should not be overlooked. Nevertheless, successful fulfillment of the federal role will be essential to facilitate state, local, and private-sector initiatives in public health, service delivery, and research.

THE CHALLENGE OF IDENTIFYING HEALTH DISPARITIES

The United States is becoming increasingly diverse. White Americans currently constitute nearly 70 percent of the population. However, by 2050, persons of color will make up nearly half of the population. In some states, such as California, this transition has already occurred, and the proportion of California's population that is Hispanic is expected to grow dramatically in the next decade. Asian/Pacific Islanders still constitute only a small proportion of the U.S. population, but they currently have the largest rate of population growth (U.S. Bureau of the Census, 1990). These demographic shifts have profound implications for health and health care in this country because minority populations experience a disproportionate burden of health problems.

Overall, African Americans continue to have some of the worst health outcomes. However, discussion of health disparities among racial and ethnic minorities must move well beyond comparisons of African Americans and whites. Indeed, there is considerable variation in health status among all of the major racial and ethnic groups including whites, African Americans, Hispanics, Asian/Pacific Islanders, and Native Americans/Alaska Natives. For example, while rates of diabetes are disproportionately high among African Americans, American Indians, and Hispanics, the prevalence of diabetes among Asians is less than that for whites (National Center for Health Statistics, 2001). There also can be considerable variation within racial and ethnic subgroups. For example, although Hispanics experience lower overall mortality rates than whites, Puerto Ricans have higher infant mortality rates than whites (National Center for Health Statistics, 2000). Some racial and ethnic subgroups have increased burdens of specific diseases. For instance, Vietnamese American women have cervical cancer mortality rates many times higher than those for other Asian and white women (IOM, 1999).

At present, the sources of such disparities remain unclear, but a wide range of explanatory factors have been suggested, including sociocultural, socioeconomic, behavioral, and biological risk factors, and environmental living conditions (Robert and House, 2000; Fremont and Bird, 2000; Williams, 1999). For example, minority populations as a whole tend to have lower socioeconomic status (SES) than other groups, and low SES is associated with poorer health, independent of race or ethnicity (Gornick, 2002). It is also generally agreed that differences in access to care, including preventive services, and racial and ethnic differences in the quality of care obtained contribute to observed disparities in health. In some minority groups and subgroups the prevalence of various conditions is especially high. Thus, the benefits of improved care for these groups may be substantially more than for others.

The challenge of understanding variations in health between and among racial and ethnic groups is further heightened as more Americans are of mixed racial and ethnic backgrounds. Although only a small proportion of respondents identified themselves as belonging to more than one racial and ethnic group on the latest census, the number of individuals in this group is expected to increase. The Office of Management and Budget (OMB, 1977) has issued guidance and developed a way to bridge the changes that should help examinations of changes over time.

THE ROLE OF RACIAL AND ETHNIC DATA IN SUPPORTING THE ESSENTIAL FUNCTIONS OF THE HEALTH CARE SYSTEM

The health system serves many important functions, but for the purposes of this paper we focus on three, with a particular emphasis on the last: ensuring the *health of the population*, ensuring *equitable access* to care, and ensuring *quality of care*. Admittedly, the system does not perform optimally in any of these areas, but it performs especially poorly in each of these areas for minority racial and ethnic populations. Data on race and ethnicity are therefore essential for improving performance for each of these functions. Documenting the extent and types of problems and identifying populations at particular risk is a crucial first step to improving performance. When available, these data convey critical information to both providers and policymakers.

Ensuring the Health of the Population

Identifying Problems

The ability to provide consistent and reliable epidemiological data on the incidence and prevalence of various health conditions and related risk factors among different racial and ethnic populations is essential to ensuring the health of the population. It also supports the rationale for allocating health care resources and developing appropriate public health interventions. For example, examination of data on race and ethnicity revealed that rates of HIV infection were rising more rapidly among African Americans and Hispanics than in any other racial and ethnic group (Shapiro et al., 1999).

Targeting Risk Factors

Risks at the individual or community level such as smoking, unsafe sexual practices, or environmental exposures are irrefutable contributors to poor health outcomes. Racial and ethnic data in federally conducted health surveys such as the CDC Behavioral Risk Factor Surveillance System (BRFSS) and others enable public health officials to better characterize the distribution of such risk factors among different racial and ethnic groups and identify emerging problems.

Access to Care

Access to care is a prerequisite for entering and staying in the health care system. Available racial and ethnic data have been used to document important differences in access between racial and ethnic groups. For example, Hispanics are substantially more likely to be uninsured than whites, and African Americans are more likely to have public insurance than whites (Collins, Hall, and Neuhaus, 1999; Hoffman and Pohl, 2000). Among blacks and whites, rates of insurance were relatively constant during the 2 decades between 1977 and 1996, but during the same period the proportion of Hispanics who were uninsured increased substantially (Weinick, Zuvekas, and Cohen, 2000).

Even when minority individuals have health insurance, they are more likely to experience barriers to care and are less likely to utilize certain types of services. For example, minority patients are less likely to report having a regular source of care (Collins, Tenney, and Hughes, 2002; Doty and Ives, 2002). Conversely, they are more likely to use the emergency room or to be hospitalized for ambulatory care-sensitive conditions such as congestive heart failure (IOM, 2003). Such utilization may reflect poorer care. Finally, because minority patients overall tend to have greater actual need for services, apparently equivalent care between racial and ethnic groups may signify underutilization by minority patients if case-mix issues are not taken into account (Lurie, 2002).

Quality of Care

The IOM report *Crossing the Quality Chasm* highlighted considerable gaps between current standards of care and the quality of care that patients actually receive. These gaps were particularly pronounced for racial and ethnic minorities. Many observers now conclude that eliminating racial and ethnic disparities is a core issue in improving quality of care (IOM, 2001).

Numerous studies have documented racial and ethnic disparities in care (see IOM, 2003, for an exhaustive review). For example, using RAND/UCLA appropriateness criteria, Laouri and colleagues (1997) showed that African Americans were only half as likely to undergo a needed coronary artery bypass graft and one-fifth as likely to undergo a percutanerous transluminal coronary angioplasty. Similarly, Ayanian and colleagues (1999) have shown that African Americans with end stage renal disease were considerably less likely to receive a referral for a renal transplant than comparable whites, even when patient preference was taken into account.

Several recent studies have also shown racial and ethnic disparities in performance on Health Employer Data and Information Set (HEDIS) process measures, which are widely used in managed-care settings (Schneider, Zaslavsky, and Epstein, 2002; Fremont et al., 2002; Virnig et al., 2002). All three studies showed that black patients were substantially less likely than whites to receive indicated care such as an annual hemoglobin A_1c test in diabetics. Virnig and colleagues (2002) also documented disparities between other racial and ethnic groups. Racial and SES disparities were also observed for several intermediate outcome measures including control of lipids and hemoglobin A_1c in diabetics and blood pressure in hypertensive enrollees (Fremont et al., 2002).

An increasing number of studies have assessed disparities in quality of care by using surveys or qualitative methods to elicit patient reports about their care. Such studies have documented significant differences in how patients from different racial and ethnic groups experience the care they receive and the kinds of barriers they encounter in accessing it. For example, in a recent Commonwealth Fund study, Asian Americans were least likely to report that their doctors understand their backgrounds and values (48 percent) compared to Hispanics (61 percent), African Americans (57 percent) and whites (58 percent) (Collins, Tenney, and Hughes, 2002). Asians were also least likely to report a great deal of confidence in their doctor. Hispanics, regardless of language skills, were more likely than other patients to report having difficulty communicating with and understanding their doctor (33 percent Hispanics and 16 percent whites) (Doty and Ives, 2002). African Americans were nearly twice as likely as whites to report being treated with disrespect during a recent visit (Collins, Tenney, and Hughes, 2002). These sorts of findings have stimulated public and private provider efforts to ensure culturally competent care and to provide language-appropriate services for their patients (IOM, 2002).

USING DATA ON RACE AND ETHNICITY TO REDUCE DISPARITIES IN HEALTH AND HEALTH CARE

The discussion above has focused on ways in which available racial and ethnic data can support essential functions of the health system by identifying populations at risk for particular conditions or with special needs, and documenting disparities in access to and quality of specific types of care. However, simply collecting racial and ethnic data and describing disparities in health and health care does not automatically lead to reductions in disparities. This information needs to be used in ways that stimulate development and implementation of efforts to effectively eliminate disparities. Thus, in this section we highlight some potential uses of racial and ethnic data to promote improved health and health care in minority populations.

Refining Public Health Initiatives and Enhancing Access to Care

Knowing which racial and ethnic population groups are most at risk can help more effectively target public health efforts. For example, documentation of high rates of HIV among African Americans and Hispanics has helped stimulate the development of federal programs that target minority groups at high risk, particularly those in low-income communities (Shapiro et al., 1999). In many instances public health efforts, such as educational campaigns, may not work equally well across different racial and ethnic groups. For instance, an apparently successful mass media campaign to educate the public about the importance of placing infants on their side or back to reduce risk of sudden infant death syndrome was subsequently shown to be far less effective among black mothers than white mothers (Malloy, 1998) largely because the educational messages were not appropriately focused on black women. Such evaluations require racial and ethnic data and are essential to refining public health efforts. At the very least they can reinforce the need to tailor public health efforts to meet the needs of different racial and ethnic groups.

Racial and ethnic data can also be used to facilitate programs designed to improve access to care. For example, African Americans and Hispanics with cancer typically face substantial barriers to obtaining and completing treatment. Their rates of participation in clinical trials involving state-of-the-art treatment protocols are especially low (National Cancer Institute, 2003). Such data have prompted the development of "Patient Navigator" programs in which culturally and language-appropriate individuals with special training are matched with patients at risk to educate them about screening and prevention measures or guide them through the treatment process once they are diagnosed. Although further evaluation is needed, such programs show promise for reducing disparities in access to cancer care and outcomes (U.S. House of Representatives, 2002).

Improving the Quality of Care

An important national strategy for improving quality of care has been the promotion of accountability for quality (IOM, 2001; Berwick, 1998). Measurement and reporting are essential components of this strategy. For example, widely used quality monitoring programs such as the National Committee for Quality Assurance (NCQA) Health Employer Data and Information Set (HEDIS) have been shown to improve performance on key quality measures among participating health plans (NCQA, 1999).

Currently, however, most health plan quality improvement efforts, including the NCQA program, are not focused on reducing racial and ethnic disparities in care (Fiscella et al., 2000). Several studies have shown that managed care alone is an insufficient mechanism to eliminate disparities in care. Thus, routine reporting of widely used quality measures separately by race and ethnicity provides an excellent opportunity to identify disparities within health plans and to apply quality improvement principles to reduce them. Performance measures are typically reported as averages across all eligible patients but are not broken down by racial and ethnic group. Consequently, important disparities within and between plans are not recognized or addressed. A number of recent studies have shown that it is feasible to obtain and use information on enrollee race and ethnicity with selected HEDIS measures to detect racial and ethnic and socioeconomic disparities (Schneider, Zaslavsky, and Epstein, 2002; Virnig et al., 2002; Fremont et al., 2002; Nerenz et al., 2002).

Since the publication of the IOM's report *To Err Is Human* (2000), reducing medical errors and improving patient safety have emerged as major areas of focus for policymakers, providers, and researchers. Though research in this area is still new, early studies show differences in the extent and nature of medical errors and problems with safety experienced among patients from different racial and ethnic groups (Burstin, 1993). Thus, just as with quality of care in general, efforts to identify and eliminate problems with patient safety can be enhanced by taking into account possible racial and ethnic differences of patients at risk.

Stimulating Value Purchasing

As reflected in the forthcoming National Quality Report, many policy makers believe that encouraging consumers and employers to base purchasing decisions on quality of care will ultimately lead to better quality and lower costs. Experts believe that information about quality is most useful to consumers when the information pertains to care received by people like themselves. Thus, reporting measures of care for specific racial and ethnic subgroups can strengthen the ability of consumers—both individuals and employers who purchase care on their behalf—to rationally choose health care providers.

In addition, individuals and advocacy groups can use racial and ethnic data in negotiations with providers, health departments, and elected officials to hold them accountable for results and to develop additional programs and policies to address disparities. In this vein, such data may ultimately have uses in domains outside the personal delivery system. For example, in some communities, examining data on diabetes prevalence and outcomes for blacks and Hispanics has revealed the need for more community-based opportunities for safe exercise. Advocacy groups have used this information in working with local officials to build walking paths and recreational facilities.

Similarly, employers and other purchasers can use racial and ethnic data to ensure that they are getting good value for their premiums. For example, after learning of the IOM report on disparities, the benefits specialist at a large national employer became concerned that her workforce, which is largely minority, may not be receiving care of appropriate quality (or quantity). She now systematically queries plans during renewal negotiations and monitors actions they are taking to reduce disparities.

Understanding the Underlying Causes of Disparities and What to Do about Them

The routine collection of data on patient race and ethnicity can also help researchers disentangle factors underlying health care disparities (IOM, 2003). Recent research efforts such as the AHRQ-sponsored Excellence

Centers to Eliminate Ethnic/Racial Disparities (EXCEED) are beginning to clarify causal factors and effective interventions. However, the development of such knowledge is likely to proceed slowly without the availability of additional racial and ethnic data with which to engage a wider group of researchers. (Also see the discussion of language preference and socioeconomic factors below.) Since causal factors and effective interventions may vary across settings, the availability of such data will also be crucial to enabling quality improvement teams to identify specific factors underlying disparities and appropriate interventions in their respective organizations. Provider organizations can enhance their own efforts by sharing best practices for reducing disparities.

As previously discussed, access to care and an array of community-level factors also have important influences on health. Expanding the availability of racial and ethnic data is a clear prerequisite both to a better understanding of how such factors affect different groups and to the development and evaluations of interventions to address them.

Ensuring Compliance with Civil Rights Law

Routine monitoring of access, use of services, and key processes and outcomes of care by race and ethnicity is essential to ensuring compliance with civil rights laws and detecting evidence of discrimination. Whether these practices are intentional or not, whether they are at the level of an individual practitioner or due to system-level problems, they can produce harmful outcomes (IOM, 2003). Title VI of the Civil Rights Act of 1964 (U.S. Office for Civil Rights, 2000) and related statutes were intended to ensure that patients from different racial and ethnic subgroups have equal access to quality care. However, enforcement of these basic rights by the Office for Civil Rights and other entities is made far more difficult without standardized, readily available data on race and ethnicity to monitor the care that different subgroups receive (Smith, 1999).

LANGUAGE PREFERENCE AND SOCIOECONOMIC FACTORS

Categorizing individuals in racial and ethnic categories has helped to identify health disparities. However, many anthropologists and other social scientists view these categories as relatively crude and inaccurate tools for understanding differences between groups (IOM, 2002). Indeed, substantial variation in important characteristics between and within racial and ethnic groups may have as much or more effect on health and the quality of care than race or ethnicity per se. Two such characteristics are language preference and SES.

Language Preference

Many individuals experience language barriers ranging from no English proficiency to limited proficiency in speaking, reading, or comprehending English (IOM, 2003). For example, it is estimated that more than one in four Asian/Pacific Islanders and Hispanics live in households where no adolescent or adult speaks English "very well" (U.S. Bureau of the Census, 1990). A recent Commonwealth Fund report documents the overwhelming barriers to care faced by non-English speaking Hispanics (Collins, Tenney, and Hughes, 2002). Language barriers often vary considerably by country of origin within racial and ethnic groups. For instance, whereas less than 2 percent of Hawaiians and 15 percent of Japanese live in households where English is not spoken well, 26–42 percent of Thais, Chinese, Koreans, and Vietnamese, and more than half of Laotians, Cambodians, and Hmong live in such households (U.S. Bureau of the Census, 1990; IOM, 2003).

A number of federal regulations encourage the use of interpreters in the health care setting (U.S. DHHS, 2000a; Office for Civil Rights, 2000; Perot and Youdelman, 2001). However, language barriers continue to pose significant problems for both patients and providers. In one recent survey, 43 percent of the Hispanics living in households where Spanish was the primary language reported having difficulty communicating with and understanding their doctor (Doty and Ives, 2002). Only half of the patients who said they needed an interpreter when visiting a doctor said they always or usually got one. In many instances (43 percent), when an interpreter was available, he or she was a member of the patient's family; rarely (1 percent) was the interpreter professionally trained (Collins, Tenney, Hughes, 2002).

Many providers are acutely aware of how language barriers and other cultural differences constrain their ability to provide effective care (IOM, 2002). In a survey of Los Angeles providers who participated in care programs sponsored by the county health authority, more than 70 percent felt that language and culture are important to the care of their patients and more than half believed that their patients did not adhere to medical treatments as a result of linguistic or cultural barriers (Cho and Solis, 2001).

In sum, routine collection of information on patients' primary language or language preference is also an essential step in identifying patient subgroups where language barriers may be present and in developing culturally competent care (Betancourt, Green, and Carillo, 2002; IOM, 2002).

Collecting information on a patient's primary language is legal and authorized under Title VI of the Civil Rights Act of 1964, though few federal statutes require it (Perot and Youdelman, 2001). Some health plans already routinely collect such data. For example, Kaiser Permanente in northern California has made cultural competency a priority for several years. As part of this effort, those using Kaiser data systems for patient care (such as making an appointment) cannot do so unless the language preference field has been filled in: the system thus prompts the user to indicate the patient's language preference and the need for translator services. Information sheets in the patient's native language describing his/her condition or treatment can often also be provided.

Socioeconomic Status

SES is a multidimensional concept that reflects an individual's access to material and social resources and assets including income, wealth, and educational credentials, as well as an individual's prestige or status in society as reflected in access to consumption of goods, services, and knowledge (Krieger, Williams, and Moss, 1997). SES is measured in a variety of ways. Income and/or education measures are the most common in the United States; measures such as occupational class may be used in Europe. Regardless of how SES is measured, numerous studies have demonstrated a consistent socioeconomic gradient in which health status, and often the quality of care received, decreases with declining SES. The gradient generally persists even when individual risk factors, including being in a minority racial or ethnic group, are taken into account (Robert and House, 2000; Kaplan, Everson, and Lynch, 2000; Gornick, 2000).

Although race or ethnicity and SES may exert some independent effects, the two are often interrelated; hence, information on SES can help highlight the underlying sources of disparities in health status and care between and among different racial and ethnic groups (IOM, 2002; Wong et al., 2002). For example, education and income levels vary substantially among minority racial and ethnic groups. Among adults, Asian Americans/Pacific Islanders were most likely to have at least a high school education (83 percent), followed by African Americans (72 percent), and Hispanics (53 percent) (Bennet and Martin, 1995). Asian Americans/Pacific Islanders also had the highest median income ($55,500); that of African Americans and Hispanics was substantially lower ($33,400 and $30,400, respectively) (U.S. Bureau of the Census, 2001). However, although Asian Americans/Pacific Islanders tend to have the highest education and income levels overall, SES varies considerably between Asian subgroups.

Although there is wide consensus that SES plays a major role in health disparities and should be taken into account whenever possible, exploring the links between SES and health disparities is a relatively new field of research. Many key issues remain unresolved—in particular the best ways to measure SES (Gornick, 2000). Consider education, one of the most widely used measures. It is popular because it is easy to measure and collect, it applies to persons who are not active in the labor force (e.g., unemployed, retired), and it is relatively stable over an adult's life span, regardless of changes in health. However, the economic and health correlates of educational level may vary by age, birth cohort, gender, class position, and perhaps most importantly, by race or ethnicity (Krieger, Williams, and Moss, 1997). For example, economic returns for a given educational level are larger for whites than for African Americans, Hispanics, and American Indians (U.S. Bureau of the Census, 1991). College-educated blacks are four times more likely than their white counterparts to become unemployed and experience consequent drops in income (Wilhelm, 1987). The limitations of individual measures have led experts to encourage simultaneous use of multiple measures to assess SES (Krieger, Williams, and Moss, 1997; Robert and House, 2000).

PUBLIC AND PRIVATE-SECTOR USERS OF RACIAL AND ETHNIC DATA

As suggested above, a variety of stakeholders have considerable interest in racial and ethnic data on health and health care. In this section, we focus on public- and private-sector entities that are likely to use such data.

Public-Sector Users

The following federal and state agencies, which are tasked with the primary responsibility to deliver or purchase health care need racial and ethnic data to ensure that care of similar quality is being provided to all users.

Center for Medicare and Medicaid Services (CMS)

The most obvious federal user of racial and ethnic data is CMS. Monitoring and reporting quality of care has been a long-standing function of the Medicare program. Research reports and other studies of the Medicare program have examined differences in utilization of Medicare services according to race. Until recently, data were available and sufficiently reliable to report about care for blacks and whites only. Improvements to the system through which Medicare receives data about race and ethnicity have made it possible to examine differences in utilization for major racial and ethnic groups, although these data still have many limitations (Lauderdale and Goldberg, 1996; Arday et al., 2000).

Such analyses have revealed important differences in the extent and nature of disparities for different minority subgroups and the need to tailor interventions to specific subgroups. Nevertheless, CMS does not yet report quality of care measures for all of the main racial and ethnic subgroups. As we discuss below, CMS's ability to accurately gauge such disparities and effectively target interventions to meet the needs of its beneficiaries is undermined by problems with existing data sources.

CMS shares responsibility for Medicaid with state governments. Because nearly 50 percent of Medicaid beneficiaries come from racial and ethnic minority groups, improving quality of care in the Medicaid program is likely to have significant advantages for beneficiaries. Geppert and colleagues in this volume (see Appendix E) discuss the state government as a user of racial and ethnic data in more detail.

Health Resources and Services Administration (HRSA)

HRSA finances delivery of care through community health centers. Unlike Medicare, HRSA collects data on the race and ethnicity of community health center users and has reported quality of care measures by race and ethnicity.

Substance Abuse and Mental Health Services Administration (SAMHSA)

SAMHSA does not directly provide care for individuals with substance abuse and mental health problems. However, it does have responsibilities for ensuring that substance abuse services are available to those in need. Fulfilling this function requires knowing the racial and ethnic characteristics of the population in need so that appropriate programs can be made available.

Department of Veterans Affairs (VA) and Department of Defense (DOD)

The VA and DOD also directly operate health care programs and have responsibilities for providing high-quality care to enrollees. The DOD has ready access to data on race and ethnicity and has conducted a number of focused studies to ascertain whether there are racial or ethnic disparities in care. The VA has had to rely largely on Social Security Administration data to obtain racial and ethnic information and until recently was able to examine quality for whites and blacks only. A recent effort to collect data about satisfaction and health status from all VA users has generated self-reported racial and ethnic data that can be used to monitor and report on quality of care. Health care providers in VA and DOD facilities are increasingly held accountable for performance on standard quality indicators but are not routinely assessed for how their performance varies between patients in different racial and ethnic groups.

Other DHHS Agencies

In addition to those federal agencies involved in the direct provision of care, other agencies of DHHS have a responsibility to ensure the equitable distribution of health care assets across the population. This includes the Agency for Healthcare Research and Quality (AHRQ), which has responsibility for monitoring health insurance and medical expenditures, and the Centers for Disease Control and Prevention (CDC), which often needs such information to fulfill public health functions. Key public health functions that require the collection of racial and ethnic data include surveillance, prevention, including immunization and health education, and communication.

Federal Employee Health Benefit Program (FEHB)

The FEHB purchases health insurance on behalf of federal employees and their dependents. As a prudent purchaser, the program has a responsibility to create conditions that ensure high-quality care for its enrollees. It also has an interest in ensuring that, as a steward of federal resources, it is getting value for the health care dollar, and that groups for which it pays the same premium actually receive equivalent services. It is highly likely, however, that FEHB cannot completely meet either of those conditions because the ability to measure quality of care for different population groups does not currently exist.

Private-Sector Users

Demands for accountability may be even more pronounced in the private sector than in the public sector. The groups most affected by this trend and most in need of data on race and ethnicity are insurers and providers. As discussed in an earlier section, employers and individual consumers also are increasingly interested in this information.

Private Health Insurers

Insurers in the managed-care and fee-for-service sectors are increasingly held accountable for providing high-quality care. Many large employers have begun requiring measures of quality, such as through HEDIS. However, lack of racial and ethnic data in the private insurance market has largely prevented assessment of quality of care for different racial and ethnic populations, let alone holding insurers accountable for providing care of similar quality. Private purchasers have interests similar to those of public purchasers in this regard. Yet they are unable to ensure that they are getting good value for all of their employees. Their inability to do so has broad implications for firms. Because good health is essential for a productive workforce, employers with large minority workforces may inadvertently undermine their organization's productivity by contracting with insurers or health plans that have large disparities in the quality of the care they provide.

Health Care Providers

Increasingly, health care providers such as hospitals, clinics, and physician practices are also being held accountable for providing high-quality care. But as previously discussed, achieving accountability is impossible without adequate data. One additional point is worth noting in this area. The IOM report *Unequal Treatment* (IOM, 2003) suggests that subconscious biases and other institutional processes may contribute to disparities in care. Without data on the existence and extent of these disparities at an institutional or practice level, it is impossible for providers—be they individuals or organizations—to know whether they have such problems and whether they are making progress in addressing them.

APPROACHES TO OBTAINING DATA ON RACE AND ETHNICITY

A variety of existing data sources contain information about individuals' race and ethnicity that is linked, or can be linked, to information about their health and health care. Such data sources have been crucial to documenting disparities in health and care for minority and low-income individuals in a variety of settings and circumstances.

Unfortunately, existing data sources, which are often targeted to specific populations, tell us little about large segments of the population and are undermined by inaccurate coding of race and ethnicity or lack of comparable measures of race and ethnicity. Below we describe some existing data sources and note limitations to their use. We focus primarily on Medicare data; descriptions of other data sources are available elsewhere (IOM, 2003; Bierman et al., 2002; U.S. Office for Civil Rights, 2001).

Medicare Data

With approximately 40 million beneficiaries, 15 percent of whom are minority, Medicare files should, in theory, be a source of racial and ethnic information on a significant portion of the population. In practice, however, the usefulness of these data has been limited by problems with their accuracy and completeness. Information on the race and ethnicity of Medicare beneficiaries is contained in the CMS enrollment database (EDB).

CMS does not actually collect racial and ethnic information directly; instead it uses entitlement information collected by the Social Security Administration on a voluntary basis from individuals applying for a new or replacement social security card. This information is then transferred to a file known as the master beneficiary record file (MBR), which, in turn, is used to populate the racial and ethnic categories in the EDB (Arday et al., 2000).

The MBR includes only four racial and ethnic categories: white, black, other, and unknown. Most experts consider Medicare data on race and ethnicity to be useful for comparing blacks vs. whites (or nonblacks); however, its usefulness for other racial and ethnic categories is more limited. The accuracy of racial and ethnic codes in the EDB may be further undermined by the fact that the EDB assigns the race of the primary wage earner to the approximately 20 percent of beneficiaries entitled to Medicare because of their association with the wage earner.

Beginning in 1994, CMS undertook a number of steps to improve the accuracy and completeness of racial and ethnic coding in the EDB. The main change was that CMS began to periodically update the EDB using another SSA file, known as NUMIDENT, that added three categories of new race and ethnicity: Hispanic, Asian American or Pacific Islander, and North American Indian or Alaska Native. Unfortunately, the SSA opted to ask applicants whether they were Hispanic in a single question about race and ethnicity rather than separately from the question about racial categories. Consequently, Hispanics were effectively coded as race rather than ethnicity (Lauderdale and Goldberg, 1996). In addition to NUMIDENT updates, CMS was able to further correct the coding in the EDB by conducting a direct-mail survey to over 2 million beneficiaries whose race was listed as "unknown or other" or who had a Hispanic surname or country of birth. CMS has also begun to work with the Indian Health Service to correct misclassification of Native American beneficiaries.

Although these improvements have substantially increased the accuracy and completeness of racial and ethnic codes in the CMS EDB, several recent analyses indicate that the accuracy of the codes is still less than 60 percent for Hispanics, Asians, and Native Americans (Perot and Youdelman, 2001). Moreover, starting in 1988 the SSA began assigning social security numbers at birth without collecting racial and ethnic data, consequently, NUMIDENT files may not include racial or ethnic data on most individuals born in the United States after 1988.

CMS also monitors the performance of the Medicare program through both Consumer Assessment of Health Plans Survey (CAHPS) and the Medicare Current Beneficiary Survey (MCBS). Fortunately, both of these surveys collect self-reported data on race and ethnicity, so accuracy and completeness are not major problems. However, the ability to analyze important subgroup responses in both of these surveys is limited by sample size considerations. If data on race and ethnicity were universal in Medicare data systems, it would be possible to oversample populations of interest when CAHPS and MCBS are administered.

Imputing Data on Race, Ethnicity, and SES

Even if the standardized collection data of on race, ethnicity, and SES were required, widespread implementation and the collection of complete data on all patients would likely take several years. In the meantime, it may be possible to impute measures of race, ethnicity, and SES from existing data sources when other sources are unavailable. The two most widely used methods of imputation for this are geocoding and surname analysis. Both of these approaches are relatively easy to perform; however, both may have limited usefulness for identifying members of some racial or ethnic subgroups.

Geocoding

This is a well-validated technique in which an individual's address is linked to census data about the area in which he or she lives in order to estimate the individual's racial, ethnic, or socioeconomic characteristics (Krieger, Williams, and Moss, 1997). Geocoding can be performed easily using commercially available software or by contracting with a vendor. In general, the smaller the census area considered, the better the estimate of an individual's characteristics. For example, Zip Codes span relatively large geographic areas containing upwards of 30,000 people, and they are not typically homogeneous in sociodemographic make-up. In contrast, census block groups averaging 1,000 residents or fewer correspond to the size of a small neighborhood and are generally quite homogeneous.

Geocoding is considered to be more useful for imputing socioeconomic characteristics of individuals based on the SES characteristics of their neighborhoods than for identifying an individual's race (Krieger, Williams, Moss, 1997). Nevertheless, geocoding can provide reliable estimates of racial characteristics among certain minority groups that tend to live in segregated neighborhoods. For example, in one recent study, geocoded measures of black race showed high concordance with individual-level measures, and analyses of disparities in quality of care in managed-care plans had essentially identical outcomes whether the geocoded or individual measure of race was used (Fremont et al., 2002). Nevertheless, geocoding may produce relatively high rates of misclassification of race or ethnicity when applied to geographic areas where racial or ethnic segregation is low for the group(s) of interest.

Surname Analysis

This technique uses an individual's last name to estimate the likelihood that they belong to a particular racial or ethnic group. The technique has been applied in a variety of ways such as to identify local concentrations of ethnic groups or to estimate the completeness of racial or ethnic identification when information is incompletely recorded (Abrahamse, Morrison, and Bolton, 1994; Lauderdale and Kestenbaum, 2000). The use of surname analysis has been limited to identifying Hispanics and Asian Americans, for which there are well-developed surname dictionaries. For example, in a recent sponsored project by the Commonwealth Fund, selected health plans used only surname analysis to identify Hispanic plan members; and found significant disparities in care between Hispanics and non-Hispanics (Nerenz et al., 2002).

Use of surname analysis for other racial or ethnic groups is much more problematic than for Hispanics and Asians and is generally not done. For instance, many Native Americans have names given to them by settlers and would be frequently misclassified as non-Hispanic whites using this technique. Depending on the name and context, misclassification can also be a significant problem even among Hispanics and Asians (Abrahamse, Morrison, and Bolton, 1994). The utility of surname analysis may increase over the next several years as new dictionaries and techniques (e.g., incorporating information from geocoding) are made available that make it possible to more accurately identify racial and ethnic subgroups among Asians (Lauderdale and Kestenbaum, 2000) and Hispanics (Peter Morrison, oral communication, August 2002).

CHALLENGES TO COLLECTING RACIAL, ETHNIC, AND SES DATA

The potential benefits of collecting racial, ethnic, and SES data are clear to many. However, data collection raises a number of concerns that must be addressed. These include methodological considerations, patient privacy and confidentiality, civil rights laws, and burdens that collecting additional data will create for various entities.

Methodological Considerations

The most appropriate approach to obtaining data depends on the specific circumstances in which the data are collected; multiple methods will probably be needed until self-reported racial and ethnic data are collected in a standardized fashion (e.g., using OMB racial and ethnic categories) and are more widely available. For example, because most private insurers do not now collect data on race or ethnicity, a mix of methods may be needed to bridge to the goal of achieving complete and accurate information over the next several years. Collecting data by self-report at or shortly after enrollment as part of an intake process is likely to provide the most accurate data for those who respond. Such an approach could be merged with imputed data, such as estimate based on geocoding, until such time as primarily collected data are available for all enrollees.

How long it might take to achieve totally self-reported data probably depends on the nature of the enrolled population and on market dynamics, including disenrollment patterns and the number of plans in an area. Collecting additional data during patient encounters could conceivably fill some important gaps, but this approach is likely to yield data of inconsistent quality and to be time consuming and impractical, especially considering the challenge of collecting data on enrollees who do not obtain care regularly. Surveys sent to all enrollees for whom racial or ethnic data are missing can help to fill gaps, as has been the case for Medicare. However, plans will need a clear reason to undertake such an effort. Their willingness to do so may depend on external pressures for reporting about quality and on consumer/purchaser demand for information.

Concerns about Privacy and Confidentiality

Many patients and consumer advocates view collecting information about an individual's race, ethnicity, and socioeconomic characteristics as intrusive and a potential invasion of privacy. Thus, ensuring that collected information remains confidential is crucial if routine collection is to occur. The passage of the Health Insurance Portability and Accountability Act (HIPPA), and the related privacy rule should go a long way to ensure that individually identifiable information about a patient's race, ethnicity, and SES is not shared inappropriately.

Nevertheless, the HIPAA Privacy Rule's primary focus is to protect personal health information rather than information about race, ethnicity, and SES. In addition, covered entities have considerable latitude in how information in medical records is used within their organization (i.e., for treatment, payment, and routine operations) as well as to which business associates to disclose information. Thus, concerns among patients, consumer advocates, and others about the privacy and confidentiality of racial and ethnic data are likely to persist.

Exposure to Civil Rights Litigation

Increasing attention to patient privacy and the financial and criminal penalties associated with violating the Privacy Rule have also likely heightened providers' concerns that routine collection of racial and ethnic data may violate civil rights laws. Indeed, a widespread misconception held by many providers and health plans is that collecting racial and ethnic data violates federal and state civil rights law (Nerenz et al., 2002; IOM, 2003; Bierman et al., 2002). A recent review of federal laws revealed no laws that prohibit the collection of racial and ethnic data (including language preference). Rather, there are numerous laws and program-specific statutes already in effect or scheduled to go into effect within the next several years that encourage or require the collection of racial and ethnic data (Perot and Youdelman, 2001). Similarly, reviews of state laws have shown that only four states have any sort of restrictions on collecting such racial or ethnic data, and these varied as to whether the restriction applied before, during, or after enrollment. At least one state required health plans to maintain information on enrollee's race and ethnicity (Perez and Satcher, 2001; Bierman et al., 2002).

Previous reviews have not explicitly addressed the legality of collecting data other than race and ethnicity that relate to an individual's socioeconomic status. However, many believe that the collection of such data for the purposes of monitoring and improving quality is consistent with civil rights laws designed to protect patients from differential treatment or discrimination based on personal characteristics.

Some plans and providers report concerns that the routine collection of racial and ethnic data and the reporting of various performance measures by race and ethnicity place them at substantially increased risk of class action lawsuits were disparities in their plans to become apparent. While this is a significant concern, it is not clear that the threat of litigation is greater when such data are collected than when they are not.

Two examples illustrate the ambivalence of health plans about the risks of collecting data on race and ethnicity. One health plan executive, in explaining why his plan was "a little gun shy," described a case in which an individual had voluntarily provided racial data at the time of application for an individually underwritten policy. Although the insurer is adamant that the individual policy was denied on the basis of multiple preexisting conditions, the individual sued the plan on the basis of racial discrimination. By contrast, an executive of a different health plan expressed his firm conviction that the act of examining quality for different racial and ethnic groups was evidence of his company's taking action to prevent discrimination, and would "probably immunize" the company against a successful civil rights lawsuit. Although some potential legal exposure could be avoided by limiting data collection until after a coverage decision has been made, there is probably no way to fully guard against legal action.

Burden of Collecting Data

In addition to concerns about patient privacy and civil rights, the entities that collect and analyze information about racial, ethnic, and SES characteristics of persons they serve may face significant costs (Nerenz et al., 2002). Adding data elements to large administrative data sets maintained by health plans can be expensive in terms of time and resources spent reconfiguring files and forms. In addition, depending on how data are obtained (e.g., at time of enrollment or visit, surveys of existing members), plans may need to devote substantial resources to actually filling in the data fields.

Once data are collected, there can be additional costs associated with analyzing and reporting measures of care and performance stratified by race, ethnicity, and SES. For example, NCQA HEDIS measures that require chart abstraction can be expensive for plans. To keep costs to a minimum the NCQA requires plans to obtain data only for a sample of approximately 400 enrollees. However, these sample sizes are not sufficient to conduct meaningful analyses for minority subgroups. Consequently, plans would need to oversample these groups (or sample larger overall populations) in order to have sufficient power to detect racial, ethnic, or SES differences. Small sample size is less of a problem for HEDIS measures that can be calculated solely from administrative data and that focus on care for common conditions such as hemoglobin A_1c checks in diabetics (Fremont et al., 2002).

Health plans have already invested substantial time and resources to reconfigure their data systems to meet HIPAA requirements. Unfortunately, although race and ethnicity are included as optional "situational" data fields for some types of patients and standardized forms, they are listed as "not used" or are not allowed on others. Consequently, routine collection of standardized racial and ethnic data for many types of enrollees is hindered rather than facilitated under current HIPAA rules (see the Bocchino paper in Appendix G for a detailed discussion).

Finally, in addition to the risk of litigation because of perceived misuse of racial or ethnic data (discussed above), some plans fear suffering business losses if minority populations they serve view the collection of such data as an effort to ration care. Such populations may also be reluctant to enroll in plans if publicly released "report cards" suggest poorer health and outcomes among minority or low-income groups, although the differences may reflect case mix and the effects of poverty rather than lower-quality care. Without either statistical adjustments for case mix differences, which are extremely difficult to present in report card format, or education of consumers, these report cards could unfairly hurt plans' efforts to increase market share in minority and low-income populations (IOM, 2003; Bierman et al., 2002).

EMERGING LEADERSHIP

Although considerable challenges to collecting racial and ethnic data remain, we believe there are reasons for optimism. One such reason is the emergence of innovative initiatives in the private sector. For example, Aetna, a large national insurer, has undertaken an extensive minority health initiative. There appear to be two related reasons for the company's decision. First, after examining U.S. demographic trends, Aetna has concluded that an increasing proportion of its members will be minorities and, therefore, both wants to and has to measure and ensure high-quality care for these members. Second, Aetna's leadership has articulated a moral imperative to act on the IOM report (2003), and to move toward ensuring that care for its members is not different on the basis of race or ethnicity.

Aetna has adopted a multipronged approach. First, it has begun collecting data on the race and ethnicity of its members (at or after the time of enrollment, in all states in which it is legal to do so) with the goal of using quality of care algorithms to measure quality for different racial and ethnic groups. Second, it is strengthening efforts to ensure that minority providers are part of its provider networks. Finally, the company is altering its marketing strategies, using a more diverse workforce, and stressing its commitment to cultural competence.

Although much of this effort is relatively recent, the insurer's staff report little, if any, consumer resistance to voluntarily providing racial and ethnic information. In fact, they have received significant positive feedback and expressions of appreciation from consumers. When asked about potential concerns about respondent burden in providing racial and ethnic data, staff stressed that such information was obtained during routine assessments of health status and needs, and that the major burden was "having to change all of our forms" and the software programs to read them.

Several other plans have reported collaboration with their state Medicaid agencies to examine racial and ethnic data for their Medicaid enrollees. The federal Medicaid Managed Care regulations encourage data sharing, and these plans report that use of these data has already led to quality improvements, particularly in the areas of diabetes and depression care.

CONCLUSIONS

Improving and broadening the collection of racial and ethnic data are critical to fulfilling core goals of the U.S. health system. While such efforts will pose significant challenges and entail real costs, ignoring this important need may ultimately be higher than definitively solving the problem.

- There is growing consensus that racial and ethnic disparities in health care reflect serious problems in the overall quality of care that may affect any patient. In this respect, failure to collect racial and ethnic data will hamper much-needed efforts to improve the quality of care, and exacerbate the erosion of trust in the consistency and quality of U.S. health care (IOM, 2003).

• Without accurate information on the racial and ethnic characteristics of populations served, public health efforts and associated resources are likely to be poorly targeted and may miss large segments of the populations most in need. Poorly managed chronic conditions or undiagnosed disease can result in more severe disease, worse outcomes, and higher health costs.

• Since minority populations make up an increasing proportion of the workforce, their unmet health needs can substantially reduce workers' quality of life and productivity, which in turn can affect the economy at all levels.

• Minority populations constitute an increasingly large segment of the health care market; thus lack of data about their care will undermine efforts to support consumer choice and stimulate market forces.

• The inability to routinely monitor the health needs and quality of care received by minority populations violates existing federal statutes and contradicts fundamental values of equity and fairness in this country. DHHS has already been sued for not requiring providers to collect and report uniform data on race and ethnicity. Without further initiatives to collect these data, additional lawsuits directed at government agencies and private health plans are likely.

Collection of reliable data on race and ethnicity is feasible, and there are good examples of its current collection and use in both the public and private sectors. The federal government can take a leadership role by ensuring the universal, ongoing collection of data in the Medicare program, and can stimulate its use in Medicaid programs and the private sector. Although private-sector data collection will likely remain voluntary, both insurers and employers have already demonstrated leadership in making the case for such data, and will hopefully set new industry standards. We anticipate that their efforts will be further encouraged by federal action in the Medicare program and increased consumer demand for such information.

The ongoing debate about the uses of and appropriate measures and methods for collecting SES and language preference data is encouraging. As has been the case in the area of racial and ethnic data collection, voluntary pilot efforts will likely inform the feasibility and usefulness of such data collection.

RECOMMENDATIONS

In the long run, the completeness and accuracy of data on race, ethnicity, language preference, and SES will improve only if this information is collected for meaningful and actionable purposes. To that end, we offer recommendations in the following areas.

Uniform Standards and Training

Standards must be developed to ensure uniform collection of data at federal, state, and local levels. In addition, there is a critical need for guidance about how these data should be collected in different settings (e.g., when, how, by whom) and training for frontline personnel in how best to do this.

Recommendation A1: Create a centralized body that can provide guidance and oversight regarding standards for and collection of data. This body should propose a set of incentives for the collection of such data, as well as penalties for failing to collect data in ways that meet a minimal standard.

Recommendation A2: Provide an adequate budget to ensure the availability of such training and data collection at state and local levels.

Public Health

The collection and use of data are critical to the functioning of the public health infrastructure, and provide a basis for ensuring that public health systems are accountable to different populations and communities.

Recommendation B3: Ensure sampling of all major racial and ethnic groups for all major epidemiologic and health status data collection efforts funded by the federal government, including those that provide important subnational data. These include, but are not limited to the Behavioral Risk Factor Surveillance System (BRFSS), National Health Interview Survey (NHIS), Medicare Expenditure Panel Survey (MEPS), Healthcare Cost and Utilization Project (HCUP), National Health and Nutrition Examination Survey (NHANES), Consumer Assessment of Health Plans Survey (CAHPS), Medicare, Medicaid, and federal employees. This will require a commitment to oversampling smaller populations in each of these surveys, and an appropriate budget for doing so.

Recommendation B4: Ensure the collection of data on race and ethnicity for all vital records.

Health Care Delivery System

Data collection in the health care system itself provides the basis for assessing disparities in care and for benchmarking progress. While these recommendations apply to Medicare, Medicaid, and the private sector, they may also be relevant to the VA and Department of Defense health care systems.

Medicare

Recommendation C5:

(a) Update and ensure the accuracy of data on race and ethnicity for current and future Medicare beneficiaries. This can be accomplished in a variety of ways, including geocoding and merging other self-reported racial and ethnic data, such as from CAHPS, into the EBD.

(b) Continue progress on identification of Native Americans. Extend assignment of Native American status beyond those living on reservations to urban dwellers.

(c) Eliminate dependency on the Social Security Administration for racial and ethnic data, or create a permanent fix to the problem of incomplete and inaccurate information. Current estimates of how much CMS spends on this range from a few million to several hundred million dollars annually. It is thus likely that redeployment of these funds would be sufficient to implement a permanent solution. In addition, steps need to be taken *now* to address the fact that data on race and ethnicity have not been collected in the social security card application process since 1988.

(d) Tie elimination of disparities to Medicare performance. The Medicare program should monitor quality of care for different racial and ethnic groups by plan, by hospital, and by state. Quality improvement organizations may be one vehicle to do this. In addition, CMS should use its purchasing power in working with Medicare+Choice plans to develop and implement data collection plans and initiatives to eliminate disparities. CMS should also consider whether the collection and reporting of accurate information about race and ethnicity should be a condition of hospital and nursing home participation in the Medicare program.

Medicaid

Recommendation C6:

(a) Extend the requirements for reporting performance for racial and ethnic groups in SCHIP to the Medicaid program.

(b) Make racial and ethnic data collection based on self-report (when feasible) mandatory for all new Medicaid beneficiaries, and update data on current beneficiaries as eligibility redeterminations are made. Require states to report these data to CMS and enforce this requirement.

(c) Provide Medicaid programs with sufficient financial support to accomplish this, and consider using financial incentives to meet this goal (such as tying a portion of the federal contribution to data completeness and accuracy).

Private Sector

The capacity of health plans and insurers, health systems, employers, providers, and consumers to use data on race and ethnicity to improve health and health care depends on availability of such data. The IOM report *Unequal Treatment* (2003) has provided further evidence for a business case for such data.

Recommendation D7: The Department of Health and Human Services, standard-setting organizations, providers and employers should take steps to modify HIPAA standards to facilitate the reporting of racial and ethnic data. This could be accomplished if DHHS and others clearly defined the business case for reporting these data. Change the accompanying guide that describes the racial and ethnic data elements for ANSI X12N 837 from a listing of "not used" to "situational" in order to facilitate reporting of racial and ethnic data.

Recommendation D8: Develop educational programs to stimulate consumerism through information provided to employers and individuals about how they can use such data.

Recommendation D9: The Department of Health and Human Services, the American Association of Health Plans, employer groups, and others should conduct education and outreach to health plans, purchasers, and employers regarding the legality of collecting data on race and ethnicity.

Research

In addition to research on disparities, expand research on data, such as types of data, relationships between variables such as race or ethnicity, SES, and education, and methods for oversampling smaller populations.

Recommendation E10: The Department of Health and Human Services should provide funding to expand research on ways to collect accurate data, including geocoding.

Recommendation E11: The Department of Health and Human Services should provide funding to expand on best ways to make self-report acceptable, training to collect data, and techniques for oversampling.

Recommendation E12: The Department of Health and Human Services and the Department of Justice should fund research to identify mechanisms that balance personal protection with protections for plans and others that use data for appropriate purposes as well as mechanisms to protect against misuse of data or detect possible redlining.

REFERENCES

Abrahamse, A.F., P.A. Morrison, and N.M. Bolton
  1994   Surname analysis for estimating local concentration of Hispanic and Asians. *Population Research and Policy Review* 13:383-398.
Arday, S.L., D.R. Arday, S. Monroe, and J. Zhang
  2000   HCFA's racial and ethnic data: Current accuracy and recent improvements. *Health Care Financial Review* 21:107-116.
Ayanian, J.Z., P.D. Cleary, J.S. Weissman, and A.M. Epstein
  1999   The effect of racial differences on patients' preferences in access to renal transplantation. *New England Journal of Medicine* 341:1661-1669.
Bennet, C.E., and B. Martin
  1995   The Asian and Pacific Islander population. In *Population Profile of the United States: 1995*. Washington, DC: U.S. Government Printing Office.
Berwick, D.M.
  1998   Crossing the boundary: Changing mental models in the service of improvement. *International Journal of Quality Health Care* 10:435-441.
Betancourt, J.R., A.R. Green, and J.E. Carrillo
  2002   *Cultural Competence in Health Care: Emerging Frameworks and Practical Approaches.* New York: The Commonwealth Fund.
Bierman, A.S., N. Lurie, K.S. Collins, and J.M. Eisenberg
  2002   Addressing racial and ethnic barriers to effective health care: The need for better data. *Health Affairs (Millwood)* 21:91-102.
Burstin, H.R.
  1993   Do the poor sue more? A case-control study of malpractice claims and socioeconomic status. *Journal of the American Medical Association.* 270:1697-1701.
Cho, J., and M. Solis
  2001   *Healthy Families Culture and Linguistics Resources Survey: A Physician Perspective on Their Diverse Member Population.* Los Angeles: LA Care Health Plan.
Collins, K.S., A. Hall, and C. Neuhaus
  1999   *U.S. Minority Health: A Chartbook.* New York: The Commonwealth Fund.
Collins, K.S., K. Tenney, and D.L. Hughes
  2002   *Quality of Health Care for African Americans: Findings from The Commonwealth Fund 2001 Health Care Quality Survey.* New York: The Commonwealth Fund.
Doty, M.M., and B.L. Ives
  2002   *Quality of Health Care for Hispanic Populations: Findings from The Commonwealth Fund 2001 Health Care Quality Survey.* New York: The Commonwealth Fund.

Fiscella, K., P. Franks, M.R. Gold, and C.M. Clancy
  2000 Inequality in quality: Addressing socioeconomic, racial, and ethnic disparities in health care. *Journal of the American Medical Association* 283:2579-2584.
Fremont, A.M., and C.E. Bird
  2000 Social and psychological factors, physiological processes, and physical health. In *Handbook of Medical Sociology*, C.E. Bird, P.C. Conrad, and A.M. Fremont, eds. Englewood Cliffs, NJ: Prentice Hall.
Fremont, A.M., S. Wickstrom, C.E. Bird et al.
  2002 *Socioeconomic, Racial/Ethnic, and Gender Differences in Quality and Outcomes of Care as it Relates to Cardiovascular Disease.* Rockville, MD: Agency for Healthcare Research and Quality.
Gornick, M.E.
  2002 Measuring the effects of socioeconomic status on health care. Pp. 45-74 in *Guidance for the National Healthcare Disparities Report*, E.K. Swift, ed. Washington, DC: National Academy Press.
  2000 Disparities in Medicare services: Potential causes, plausible explanations, and recommendations. *Health Care Financial Review* 21:23-43.
Hoffman, C., and M. Pohl
  2000 *Health Insurance Coverage in America: 1999 Data Update.* Washington, DC: The Kaiser Commission on Medicaid and the Uninsured.
Institute of Medicine
  2003 *Unequal Treatment: Confronting Racial and Ethnic Disparities in Health Care.* B.D. Smedley, A.Y. Stith, and A.R. Nelson, editors. Committee on Understanding and Eliminating Racial and Ethnic Disparities in Health Care. Board on Health Sciences Policy. Washington, DC: The National Academies Press.
  2002 *Speaking of Health: Assessing Health Communication Strategies for Diverse Populations.* Committee on Communication for Behavior Change in the 21st Century: Improving the Health of Diverse Populations, Board on Neuroscience and Behavioral Health. Washington, DC: The National Academies Press.
  2001 *Crossing the Quality Chasm: A New Health System for the 21st Century.* Committee on Quality of Health Care in America. Washington, DC: National Academy Press.
  2000 *To Err Is Human: Building a Safer Health System.* Linda T. Kohn, Janet M. Corrigan, and Molla S. Donaldson, Editors. Committee on Quality of Health Care in America. Washington, DC: National Academy Press.
  1999 *The Unequal Burden of Cancer: An Assessment of NIH Programs and Research forMinorities and the Medically Underserved.* M.A. Hayes and B.D. Smedley, eds. Washington, DC: National Academy Press.
Kaplan, G.A., S.A. Everson, and J.W Lynch
  2000 The contribution of social and behavioral research to an understanding of the distribution of disease: A multilevel approach. In *Promoting Health: Intervention Strategies from Social and Behavioral Research*, B.D. Smedley and S.L. Syme, eds. Washington, DC: National Academy Press.
Krieger, N., D.R. Williams, and N.E. Moss
  1997 Measuring social class in U.S. public health research: Concepts, methodologies, and guidelines. *Annual Review of Public Health* 18:341-378.
Laouri M., R.L. Kravitz, W.J. French et al.
  1997 Underuse of coronary revascularization procedures: Application of a clinical method. *Journal of American College of Cardiologists* 29:891-897.

Lauderdale, D.S., and J. Goldberg
   1996   The expanded racial and ethnic codes in the Medicare data files: Their complete-
          ness of coverage and accuracy. *American Journal of Public Health* 86:712-716.
Lauderdale, D.S., and B. Kestenbaum
   2000   Asian American ethnic identification by surname. *Population Research and Policy
          Review* 19:283-300.
Lurie, N.
   2002   Measuring disparities in access to care. Pp. 99-147 in Institute of Medicine, *Guid-
          ance for the National Healthcare Disparities Report*, E.K. Swift, ed. Washington,
          DC: The National Academies Press.
Malloy, M.H.
   1998   Effectively delivering the message on infant sleep position. *Journal of the American
          Medical Association* 280:373-374.
National Cancer Institute
   2003   *Plans and Priorities for Cancer Research*. Bethesda, MD: National Cancer Institute.
National Center for Health Statistics
   2001   *Health, United States, 2000, with Socioeconomic Status and Health Chartbook*.
          Hyattsville, MD: National Center for Health Statistics.
   2000   *Health, United States, 1999, with Socioeconomic Status and Health Chartbook*.
          Hyattsville, MD: National Center for Health Statistics.
National Committee for Quality Assurance
   1999   *HEDIS 2000: Narrative—What's in it and Why it Matters*. Washington, DC:
          National Committee for Quality Assurance.
Nerenz, D.R., V.L. Bonham, R. Green-Weir, C. Joseph, and M. Gunter
   2002   Eliminating racial/ethnic disparities in health care: Can health plans generate reports?
          Health Affairs (Millwood) 21:259-263.
Office of Management and Budget
   1977   *Racial and Ethnic Standards for Federal Statistics and Administrative Reporting*.
          (Statistical Directive 15). Washington, DC: Office of Management and Budget.
Perez, T.E., and D. Satcher
   2001   Letter to Dr. Bruce Zimmerman, President, American Diabetes Association, from
          T.E. Perez, Director, Office for Civil Rights, U.S. DHHS, and D. Satcher, Assistant
          Secretary for Health and Surgeon General, U.S. DHHS, January 19, 2001.
Perot, R.T., and M. Youdelman
   2001   *Racial, Ethnic, and Primary Language Data Collection in the Health Care System:
          An Assessment of Federal Policies and Practices*. New York: The Commonwealth
          Fund.
Robert, S.A., and R. House
   2000   Socioeconomic inequalities in health: An enduring social problem. In *Handbook of
          Medical Sociology*, C.E. Bird, P.C. Conrad, and A.M. Fremont, eds. Englewood
          Cliffs, NJ: Prentice Hall.
Schneider, E.C., A.M. Zaslavsky, and A.M. Epstein
   2002   Racial disparities in the quality of care for enrollees in medicare managed care.
          Journal of the American Medical Association 287:1288-1294.
Shapiro, M.F., S.C. Morton, D.F. McCaffrey et al.
   1999   Variations in the care of HIV-infected adults in the United States: Results from the
          HIV Cost and Services Utilization Study. *Journal of the American Medical Associa-
          tion* 281:2305-2315.
Smith, D.B.
   1999   *Health Care Divided: Race and Healing a Nation*. Ann Arbor: University of Michi-
          gan Press.

U.S. Bureau of the Census
  1991   *Current Population.* Washington DC: U.S. Bureau of the Census.
  1990   *1990 Census of Population: Asians and Pacific Islanders in the United States.* Washington, DC: U.S. Bureau of the Census.
U.S. Department of Health and Human Services
  2000a *Healthy People 2010: Understanding and Improving Health.* Washington, DC: U.S. Government Printing Office.
  2000b *Healthy People 2010.* Washington, DC: U.S. Department of Health and Human Services.
U.S. Office for Civil Rights
  2001   *HHS Directory of Health and Human Services Data Resources.* Washington, DC: U.S. Office for Civil Rights.
  2000   *Title VI of the Civil Rights Act of 1964: Policy Guidance on the Prohibition Against National Origin Discrimination as It Affects Persons with Limited English Proficiency.* Washington, DC: U.S. Office for Civil Rights.
U.S. House of Representatives
  2002   H.R. 5187: Patient Navigator, Outreach, and Chronic Disease Prevention Act of 2002. 107th Congress, 2d session.
Virnig, B.A., N. Lurie, Z. Huang, D. Musgrave, A.M. McBean, and B. Dowd
  2002   Racial variation in quality of care among Medicare+Choice enrollees. *Health Affairs (Millwood)* 21(6):224-230.
Weinick, R.M., S.H. Zuvekas, and J.W. Cohen
  2000   Racial and ethnic differences in access to and use of health care services, 1977 to 1996. *Medical Care Research Review* 57:36-54.
Wilhelm, S.M.
  1987   Economic demise of blacks in America: A prelude to genocide? *Journal of Black Studies* 17:201-254.
Williams, D.R.
  1999   Race, socioeconomic status, and health. The added effects of racism and discrimination. *Annals of the New York Academy of Sciences* 896:173-188.
Wong, M.D., M.F. Shapiro, W.J. Boscardin, and S.L. Ettner
  2002   Contribution of major diseases to disparities in mortality. *New England Journal of Medicine* 347:1585-1592.

Appendix E

State Collection of Racial and Ethnic Data

Jeffery J. Geppert, Sara J. Singer, Jay Buechner, Lorin Ranbom, Walter Suarez, and Wu Xu*

INTRODUCTION

The purpose of this paper is to analyze variations in current methods for the collection of racial and ethnic data among states and between states and the federal government, to analyze the costs and benefits of enhancing or standardizing such data collection, and to describe selected recommended practices for the collection and use of data on race and ethnicity. States obviously have an interest in promoting the health and welfare of all their citizens and ensuring minimum health disparities among population subgroups. The ability to identify meaningful characteristics of individuals, including their race and ethnicity, for purposes of program planning, implementation, and evaluation is an important tool of public policy at the state and local government levels. Therefore, states have made significant efforts to collect detailed health data by race and ethnicity (and other cultural and socioeconomic categories) (National Conference of State Legislatures, 2002). These data are beneficial for planning program eligibility, identifying populations with particular needs, generating hypotheses about potential causes, and tracking progress. However, there are also important barriers and costs

*Jeffrey J. Geppert and Sara J. Singer are staff researcher and senior research scholar, respectively, at the Stanford University Center for Health Policy; Jay Beuchner is chief at the Rhode Island Department of Health Office of Health Statistics; Lorin Ranbom is chief at the Ohio Department of Human Services Office of Health Service Research; Walter Suarez is director of the Minnesota Health Data Institute, InformationSystems and Operations; and Wu Xu is director for the Utah Department of Health, Office of Health Care Statistics.

to collecting such data, including problems with small numbers for particular subgroups, tensions between sufficient detail and meaningful aggregation, reconciling trade-offs with data collection of other important individual characteristics, and other resource constraints.

Through interviews with health agency staff and review of published sources, we examined how selected states approach the collection of racial and ethnic data. The background section describes the most common data sources used by states that collect racial and ethnic data. The methods section describes our data collection procedure for interviews and data analysis. The results section describes some of the variation among states in the collection of data on race, ethnicity, and other socioeconomic characteristics, and the benefits and costs of such data collection from the perspective of state administrators. The conclusions section provides some considerations for further research.

BACKGROUND

Data Sources

States collect racial, ethnic, and other socioeconomic data through multiple data sources as part of their administration of state and federal health programs. Some of the most common data sources and their characteristics are described below and summarized in Table E-1.

TABLE E-1 Summary of Characteristics of State Racial and Ethnic Data Sources

Data Source	Source of Racial Ethnic Data	Self-Reported or Third-Party	Coding Standard
Hospital discharge	Providers	Self-report/ Perceptual assessment	UB-92/ State-specific
Vital statistics—death	Funeral director	Family inquiry	OMB/NCHS
Vital statistics—birth	Mother	Self-report	OMB/NCHS
Cancer registries	Providers	Self-report/ Perceptual assessment	OMB/SEER/ NPCR
Medicaid	County workers	Self-report/ Perceptual assessment	OMB
Health interview surveys	Telephone survey	Self-report	State-specific

NOTE: UB-92 = Uniform Bill for Hospitals; OMB=Office of Management and Budget; NCHS = National Center for Health Statistics; SEER = Surveillance, Epidemiology, and End Results; NPCR = National Program of Cancer Registries.

Hospital Discharge Abstracts

Some 37 states collect discharge abstracts on patients admitted to nonfederal hospitals (that is, excluding VA and Department of Defense facilities). Approximately 85 percent of the states use a Uniform Bill for Hospitals (UB-92) format for collecting discharge data (AHRQ, 1999). Because the primary purpose of the UB-92 is to pay a claim, racial, ethnic, and socioeconomic data are not included in the UB-92 core billing standards. However, 27 states (55 percent) collected data on patient race and ethnicity as a part of their inpatient data using state fields.

Vital Statistics

All 50 states collect birth and death certificate data, and virtually all states provide the National Center for Health Statistics (NCHS) with racial, ethnic, and socioeconomic data from birth and death records. The standard certificates are defined by NCHS and data collection is specified in a certain way. Most states use the standard for reporting to NCHS, although there is variation at the local level. About 12 states collect data differently from the standard.

Cancer Registries

Forty-five states have cancer registries. The National Cancer Institute (NCI), through its Surveillance, Epidemiology, and End Results (SEER) program, sponsors 11 population-based registries. The Centers for Disease Control and Prevention, through its National Program of Cancer Registries (NPCR), sponsors registries where the NCI does not. Together, the two agencies cover 100 percent of the U.S. population. Many registries began in individual hospitals in accordance with guidelines established by the American College of Surgeons (ACoS) as part of the requirements for accreditation of oncology services. The main purpose of the hospital registry was to provide physicians with the data needed to maintain quality of care through peer review and to compare performance with recognized standards. Racial, ethnic, and socioeconomic characteristics are important data to epidemiologists who investigate cancer, so the categories are quite detailed. Differences in incidence rates among different racial, ethnic, and socioeconomic groups generate hypotheses for researchers to investigate.

Medicaid

All 50 states participate in the state-federal Medicaid program. Enrollment data include information on each beneficiary's race, ethnicity, and income. The data are collected from hospitals, physicians, skilled-nursing facilities, and other practitioners participating in the Medicaid program.

Health Interview Surveys

Many states collect supplementary data on self-reported health status through telephone and other surveys. Health assessment surveys based on large populations give state health planners, policymakers, county governments, and community organizations more detailed pictures of the health and health care needs of various segments of the population. Typical information includes where and how people get health care as well as the number of adults and children without health insurance. Health surveys also collect information on important health conditions such as cancer, diabetes, and asthma.

METHODS

To collect information for this paper, we interviewed and surveyed representatives in four states (Rhode Island, Ohio, Minnesota, and Utah) from various state agencies responsible for the collection of health data that include information on individual race or ethnicity. Participating states and individuals were selected nonrandomly, based on recommendations from officials in the National Association of Health Data Organizations. For five selected states (California, Florida, New York, Utah, and Wisconsin), we also examined various data sources to investigate the nature of the racial, ethnic, and socioeconomic status data actually collected. The survey instrument is included at the end of this appendix. We interviewed and collected written responses from state administrators responsible for the most common state data sources (i.e., state discharge data, vital statistics, cancer registries, Medicaid enrollment, and health interview surveys). The questions were divided into two sections. The first included questions about the perceived benefits of collecting, enhancing, and standardizing racial and ethnic data. The second section included questions about the perceived barriers or costs to enhanced or standardized racial and ethnic data collection. The questions are summarized below:

Questions on Costs
- What racial and ethnic data do your agency currently collect?
- How is this information collected? Self-reported? Reported by clinician or other third party?
- What organizations report this information to your agency?
- How are the data your agency collects currently used in your state? To whom are they reported?

- Who has the authority to make a decision about the data that your agency collects? Are data elements governed by law, regulation, agency authority, other?
- What stakeholders would be opposed to changes to the data your agency currently collects?
- How long would it take your agency to implement the data requirements? What would require the most time?
- Can you quantify or estimate the costs to your agency and the organizations reporting to your agency of any of the changes necessary to implement new data collection requirements?

Questions on Benefits
- What is the intended purpose(s) of the racial and ethnic data your agency currently collects?
- What qualitative benefits to your agency, the organizations reporting to your agency, other government entities, beneficiaries of public programs, or the public (i.e., stakeholders) do you foresee as a result of more detailed or standardized data collection? Better service provision? Greater comparability in reporting? Opportunities for improved quality of care?
- Can you identify any quantitative benefits to your agency or other stakeholders that might result from more detailed or standardized data collection? Less supplementary data collection? Opportunities for economies of scale? Simplification? Improved ability to target public or private programs?
- To whom would these benefits accrue? Your agency? Other state or local agencies? Hospitals and other health care providers? Public programs or program beneficiaries?
- How long do you expect it would take for your agency or other stakeholders to achieve these benefits? Immediate? Less than 1 year? Less than five years? Longer than 5 years?
- Can you quantify or estimate the savings or other benefits to your agency or other stakeholders due to more detailed or standardized data collection?

In addition, state representatives provided information about recommended practices they have experienced or observed. Recommended practices are strategies and practices related to the collection and use of racial and ethnic data that are, in the opinion of the state data organization representatives, particularly effective.

RESULTS

Our emphasis was not on a systematic analysis of all state collection practices of racial, ethnic, and socioeconomic status data, but rather a review of selected state practices in order to identify important issues and considerations for future research. This section describes the results from our interviews with state health administrators, a review of published documents, and preliminary data analyses. First, we provide examples of variation among and within states in the reporting of racial, ethnic, and socioeconomic status (SES) data. Second, we provide examples of variation among states in the actual collection of racial, ethnic, and SES data. Third, we highlight the perceived benefits and barriers or cost the collection of these data. Finally, we describe some recommended practices of states that have used such data, and of other cultural or socioeconomic data, to improve health and health delivery in their states.

Variation Among States in Reporting

States vary in the categories used to collect racial, ethnic, and socioeconomic status data for the common data sources described above. However, the variation is more extensive within states among the various types of data than among states for a particular data source. As noted earlier, race and ethnicity are not included among the UB-92 core elements for state hospital discharge data. Individual states have added either a single data element on race, ethnicity, and socioeconomic status or two separate elements. The actual categories of elements vary from state to state (Table E-2). In particular, states vary in the treatment of Hispanic ethnicity, as a field either separate from race or interacted with race. There is also variation in the categories for Asian and Pacific Islander, and in the treatment of unknown or no response.

There is even more variation within states, among discharge data and the other types of data that states collect. There is a spectrum of specificity of data collected: in general, hospital discharge data are at one extreme, with limited racial and ethnic categories; survey and registry data are at the other extreme, with multiple categories, including the ability to record more than one race per individual; and Medicaid enrollment data and vital statistics data are in the middle. For example, the California Cancer Registry has 29 race categories and 9 Hispanic ethnicity categories (Table E-3). In contrast, California Medicaid has 18 race categories. Birth and death records in California include 16 race categories and 7 Hispanic ethnicity categories.

One accompanying factor in the extensive collection of racial, ethnic, and SES data is the need for training in data collection. The more extensive the data collection, the more intensive the training required. Vital statistics and registries both have extensive training documentation for the reporting of racial, ethnic, and SES data. The categories of race and ethnicity were

TABLE E-2 Racial and Ethnic Data Categories for Selected State Discharge Abstracts

State	CA	FL	NY	UT	WI
Separate race/ethnicity	Y	N	Y	N	Y
Ethnic categories					
Hispanic	X		X		X
Non-Hispanic	X		X		X
Unknown	X		X		X
Racial categories					
White	X	X	X		X
Black	X	X	X		X
White Hispanic			X	X	
Black Hispanic			X	X	
White, non-Hispanic				X	
White, Hispanic				X	
Nonwhite, Hispanic					
Nonwhite, non-Hispanic					
Native American/Eskimo/Aleut	X	X	X		X
Asian/Pacific Islander	X	X			X
Asian			X		
Native Hawaiian or Other Pacific Islander			X		
Other	X	X	X		X
Unknown	X		X	X	X
No response		X		X	

NOTE: Blanks in columns indicate that the state does not use the cited category.

central to the origins of these training systems and to their current development and use.

Variation Among States in Data Collection

In addition to variation among states in the categories of data collected, states also vary in the completeness of that data. Missing data could be the result of various factors, including whether the data are formatted into a common State Inpatient Data (SID) format. The SID data use a common denominator approach to coding race and ethnicity, and variation among states in the coding may mean that some states do not have race or ethnicity reported in the Healthcare Cost and Utilization Project (HCUP) data set. States with more diverse populations might be more proficient or have more extensive training in data collection and reporting.

States also vary in their requirements for compliance by data suppliers to report racial ethnic data to the state agency maintaining the discharge data system. States may mandate the reporting of racial, ethnic, and socio-economic status data, or they may mandate the reporting of discharge data but not require race or ethnicity to be a part of the record.

Table E-4 shows that states that require the reporting of race or ethnicity as part of the discharge data submission have higher rates of compliance than states that collect the data voluntarily or do not require resubmission if the data are missing or invalid. States with mandated reporting showed 97 percent compliance in data collection and reporting, while states with voluntary reporting showed 83 percent compliance. The range of rates of compliance among hospitals was also narrower in the mandatory states

TABLE E-3a California Cancer Registry Categories of Race

White	Laotian	Samona
Black	Hmong	Tongan
American Indian, Aleutian, or Eskimo	Kampuchean (Cambodian)	Melanesian, NOS
Chinese	Thai	Fiji Islander
Japanese	Micronesian, NOS	New Guinean
Filipino	Chamorro	No Further Race Documented
Hawaiian	Guamanian, NOS	Other Asian, Including Burmese, Indonesian, Asian, NOS and Oriental, NOS
Korean	Polynesian, NOS	Pacific Islander, NOS
Asian Indian, Pakistani, Sri Lankan (Ceylonese), Nepalese, Sikkimese, Bhutanese, Bangladeshi	Tahitian	Other
Vietnamese		Unknown

NOTE: NOS = Not otherwise specified.

TABLE E-3b California Cancer Registry Categories of Hispanic Ethnicity

Non-Spanish, Non-Hispanic
Mexican (including Chicano, NOS)
Puerto Rican
Cuban
South or Central American (except Brazilian)
Other Specified Spanish Origin (includes European)
Spanish, NOS; Hispanic, NOS; Latino, NOS
Spanish Surname Only
Unknown Whether Spanish or not

NOTE: NOS = Not otherwise specified.

TABLE E-4 Compliance Rates by State Collection Directive

Discharges	Mandatory Compliance		Voluntary Compliance	
	% missing	% compliance	% missing	% compliance
Average	3.2	96.8	17.17	82.9
Minimum	7.4	92.6	76.6	23.4
Maximum	0.1	99.9	0.0	100.0

SOURCE: National Association of Health Data Organizations (NAHDO).

than in the voluntary states. The results again seem to demonstrate that reporting requirements and training matter in data collection.

Benefits of the Collection of Racial, Ethnic, and Socioeconomic Status Data

The state administrators we interviewed identified five main benefits of collecting racial, ethnic, and SES data in terms of the programs administered by their agencies.

(1) Identifying population needs

States use data on race and ethnicity to measure the health of subgroups, in terms of both levels and trends, and to identify the nature and extent of potential disparities or other special needs in health among subgroups.

(2) Program planning

States use data on the race and ethnicity of residents to plan program eligibility, especially by identifying subgroups that are growing faster than the general population.

(3) Program design

States use data on race and ethnicity to develop hypotheses about the potential causes of health disparities or other special needs among subgroups and to develop special initiatives to address those problems.

(4) Program evaluation

States use data on race and ethnicity to evaluate the effectiveness of existing initiatives targeted to addressing health disparities or other special needs among subgroups, to monitor the performance of health systems, and to track overall improvement in health.

(5) Public reporting

States use data on race and ethnicity to report to the public, health care professionals, the state legislature, and federal government on the health status of the population of state and federal program participants.

Barriers and Costs Involved in the Collection of Racial, Ethnic, and Socioeconomic Status Data

(1) Small numbers

For many states, there are relatively few individuals in particular subgroups. This is often characterized as a self-sustaining problem, as the data to document the actual numbers and to justify potential additional data collection are not available. Others also argue that from a public policy perspective, small, subgroups are even more likely to have special needs that go unrecognized and unaddressed, so that it is even more important to collect the data for these subgroups. For example, particular Asian or Hispanic ethnicities can constitute small distinct communities with distinct needs and programs. Approaches to dealing with small numbers include averaging over multiple years and reporting confidence intervals on estimates.

(2) Beyond "white vs. nonwhite"

In reporting and analysis, there are often trade-offs between using detailed racial and ethnic categories and aggregating to more general categories. Detailed categories may permit more specific inferences and hypotheses, but they add to the complexity of the analysis and may suffer from small numbers that make rates difficult to interpret. General categories may simplify the presentation and analysis, but at the expense of detail necessary to reflect local circumstances or particular conditions, or to provide sufficiently actionable information. Demand for particular detailed analyses from state policy makers and other constituencies are often the deciding factor on the level and extent of aggregation, rather than considerations based on evidence or reporting standards.

(3) Information crowd-out

In data collection, careful consideration must be paid to constraints on time and resources that may require reconciling the relative importance of additional data collection on race and ethnicity with data collection of other important individual characteristics. Surveys in particular face constraints related to respondents, and administrative data face constraints related to providers and other organizations that collect the data on any given individual.

(4) Resource constraints

Any modifications to existing data collection require resources for programming, database development, and training. In order for these costs to be justified, there must be clear supporting arguments and demonstrated need.

Recommended Practices

In order to realize these benefits and overcome these barriers and costs, states have adopted various strategies to collect and use data on race, ethnicity, and socioeconomic status. These "recommended practices" are grouped below in four general categories: data collection, training, research and analysis, and reporting.

Data Collection

Initiatives in data collection include the development of policies and procedures for the collection of racial, ethnic, and SES data, practices to ensure the quality of the data collected, and new initiatives to collect data on previously underreported subpopulations.

Policies and procedures. Examples of racial, ethnic, and SES data collection policies and procedures include agencywide policies covering some or all of the following: minimum standard racial and ethnic categories to be collected, format for data collection forms, language for telephone or self-administered surveys, recommended method for collecting data from subjects, standard data edit criteria, and standard for maximum missing responses. Some state agencies took the opportunity to develop such policies as part of the conversion to the revised federal guidance implemented with the 2000 census.

The Family Outcomes Project at the University of California-San Francisco and the Center for Health Statistics in the California Department of Health Services developed a document entitled "Guidelines on Race/Ethnicity Data Collection, Coding, and Reporting" to standardize practices across the department. The guide includes recommended practices for data collection, coding and tabulation.

Quality control. As we have seen, ensuring data completeness and quality is as important as well-defined policies and procedures. Methods of quality control include reabstraction and/or recontact studies for specific databases, routine or special feedback of summaries of collected data to collectors, and linkage studies across databases.

Initiatives to collect data on previously underreported subpopulations. Improvement of data collection on specific subpopulations requires special effort (Office of Minority Health, 2000). In one example, officials with the Ohio Commission on Minority Health partnered with the National Council of La Raza, state agencies, and community organizations to develop the first demographic overview of Ohio's Latino population, from recent history and heritage to educational attainment and health status.

Training

We have also seen that programs vary in the intensity of the training, which can be critical when the state is relying on other institutions for primary data collection. Some examples include state-sponsored training sessions or state-developed materials for specific groups of data collectors (internal or external to the agency), such as primary care physicians for reportable diseases, funeral directors for death certificates, and hospital admissions clerks for hospital discharge data and Emergency Department data. Again, such programs have been developed to support the change to the new federal guidelines.

In addition to training data collection professionals, more general cultural sensitivity training and interventions among health care providers can lead to improved data collection or measurable improvements in the health of subgroups. For example, Alabama recently trained 29 bilingual people in English-to-Spanish medical interpreter skills—including skills in medical terminology, ethics, interviewing, and Hispanic culture. The New York University Center for Immigrant Health offers training to bilingual people in medical interpretation and a pilot program in remote-simultaneous medical interpretation, where practitioners and patients wear headsets to communicate in several different languages.

Research

Data that are used are more likely to be collected. Cross-sectional studies of disparities (statewide or local), trend analyses for racial and ethnic groups, Healthy People 2000/2010 baseline and progress studies (U.S. Department of Health and Human Services, 1990, 2000), minority health data on agency websites, and web-based dissemination systems with racial and ethnic analysis options are all examples of initiatives to provide the incentive for collecting better data on race, ethnicity, and SES.

Such studies can lead to the development of initiatives to address measured health disparities. In South Carolina, surveillance studies demonstrated that the state had one of the highest rates of prostate cancer mortality in the nation. The identification of culturally related barriers to education and screening efforts was a central strategy for reducing overall mortality. The South Carolina Office of Minority Health (OMH) gathered focus groups of African American men and women and uncovered barriers to screening. These barriers included ideas about masculinity, misconceptions about how cancer develops, fear of death, and lack of knowledge about the location and function of the prostate. As a result, the OMH developed an information campaign called "Real Men Get It Checked," which has been used by hospitals and churches throughout the state.

Reporting

Examples of efforts to overcome obstacles to using data collected for reporting purposes include agencywide policies on categories to use when presenting multiple race responses and methods to allow "bridging" from previous categories to new categories.

South Carolina's Budget and Control Board linked data from state agencies and the private sector to create a fuller picture of various populations served in the state. The board's Office of Research and Statistics linked Medicaid claims, child-care vouchers, education, welfare, vocational rehabilitation, mental health services, motor vehicle crashes, juvenile justice, private inpatient hospitalizations, emergency room visits and admissions, home health visits, and other services. The data system uses unique person numbers rather than personal identifiers to ensure confidentiality.

CONCLUSIONS

States' efforts to collect racial, ethnic, and SES data either have evolved over time in response to changing circumstances, as in the case of hospital discharge abstracts and other sources of administrative data, or have been carefully planned in advance and in detail, as in the case of many of the state-based health assessment surveys. These data collection efforts meet a variety of public policy needs, and must be tailored to specific local conditions and program characteristics. Yet standardized data collection, analysis and reporting practices are more likely to produce accurate and useable information. We conclude with a few considerations for future research in meeting these multiple objectives.

A "Standardized" Set of Socioeconomic Factors

The identification of a standard set of socioeconomic and cultural factors of importance in identifying meaningful subgroups could be implemented across data sources. In addition to race and ethnicity, other individual characteristics might include income, education, and insurance status, based on existing evidence from health services and clinical research. A "minimum" set of factors could be used in administrative data systems, and a "maximum" set for registries and surveys. Defined procedures could permit aggregation from detailed categories into more general categories in two ways: vertically and horizontally. Vertical integration would allow, for example, Hawaiian, Pacific Islander, and various Asian nationalities to be pooled into a broader category, if desired. The level of detail could be tailored to specific local circumstances. Horizontal integration would include defined procedures for combining across socioeconomic and cultural factors. Evidence might suggest certain combinations or weights that might be relevant for particular uses, with race and ethnicity assuming more or less weight as appropriate.

Administrative and Nonadministrative Data

In health services research and performance measurement, there is always a perceived dichotomy between the relative merits of using administrative data and data collected from other, more detailed data sources, such as medical charts or individual surveys. Administrative data sources have the advantages of a relatively low cost to collect, universal coverage of entire populations, and transparency and reproducibility of data collection methods and analysis results. The disadvantages of administrative data are the lack of detailed information on individual demographic and clinical characteristics needed for meaningful analyses. Nonadministrative data such as medical chart abstraction and surveys can overcome these limitations, but often at considerable cost and limited availability and reproducibility.

Rather than selecting one approach over another, each approach should be pursued simultaneously and the bridge between them narrowed through standardization and automation. Standardization can enhance administrative data systems by adding additional data elements that are important for analysis of health disparities and quality of care. Standardization can also facilitate the automation of clinical data. Training is an essential and often overlooked component to this process. The lessons learned and practices developed for training in vital records and registries in particular could be applied to administrative data systems more generally.

Federal Standards and State Incentives

Federal standards and state incentives to adopt them have an important role in creating a common infrastructure for the collection of racial, ethnic, and socioeconomic status data. For vital statistics, where NCHS has adopted a standard and provides resources to states to implement those standards, states have adopted the standard and adhere to it in their own reporting. In contrast, when the decennial census changes to OMB-specified categories, states may not comply and the federal government has no control over what states do (due to state mandates controlling the collection). To change to the new census/OMB standards, states will have to go through their legislative committees and rule-making procedures, and the change rate will vary by state. Similarly, the current HCUP project, a state and federal partnership to collect hospital discharge data, relies on states to report what they have and in what format they have it. The "common denominator" data set contains a mix of categories and completeness, as we have seen. The incentive for states to change to a uniform format does not exist.

When the federal government takes the leadership role in setting standards and assisting states in paying for data collection in keeping with the standards, states are motivated and generally comply. Such standards also play an important role in building local constituencies and assisting states in moving from using the data for planning purposes only to using the data for research and analysis. As they may be adaptable to local circumstances, federal standards often make it easier to build the case for local implementation and use.

State Variation

Finally, any federal standard must permit individual tailoring to meet state requirements and to reflect local conditions. In some states and for some instruments, going beyond the standard set will be essential to understand the state's racial, ethnic, and other socioeconomic diversity.

REFERENCES

Agency for Healthcare Research and Quality
　1999　*Healthcare Cost and Utilization Project 1998 Data Availability Inventory.* (Conducted by NAHDO and Medstat). Rockville, MD: Agency for Healthcare Research and Quality.
National Conference of State Legislatures
　2002　Racial and ethnic disparities in health care. *State Lawmakers' Digest* 2(4): Summer.
Office of Minority Health
　2000　*Assessment of State Minority Health Infrastructure and Capacity to Address Issues of Health Disparity.* (Developed by COSMOS Corporation under Contract No. 282-98-00127). Washington, DC: Office of Minority Health.
U.S. Department of Health and Human Services
　2000　*Healthy People 2010: Understanding and Improving Health.* Washington, DC: U.S. Government Printing Office.
　1990　*Healthy People 2000.* Washington, DC: U.S. Department of Health and Human Services.

SURVEY INSTRUMENT

National Academy of Sciences
Panel on DHHS Collection of Race and Ethnicity Data
Cost-Benefit Analysis of State-Level Data Collection

The questions below are designed to solicit information from data collection agencies in selected states about the potential costs and benefits of enhancing and standardizing race and ethnicity data collection in the state and among states and the federal government. In answering the questions below, please think specifically about your own agency's data collection efforts and, where indicated and to the extent possible, the data collection efforts of other entities in your state. While we are interested in race, ethnicity, and socioeconomic status data in general, our analysis will focus on the following data sources: Medicaid data, Hospital Discharge Abstract data, Birth and Death Records, Cancer Registry data, State Health Interview Survey data, and Behavioral Risk Factor Surveillance System data. Therefore, we request that you also focus your responses on the subset of these data sets about which you are aware.

Please respond briefly in writing to the questions below, and return your response to Sara Singer at singer@healthpolicy.stanford.edu and Jeff Geppert at jgeppert@nber.org. A conference call will follow.

Costs: What would be the costs of enhancing and standardizing race and ethnicity data collection in your state? Nationally?

• What race and ethnicity data do your agency currently collect? Please specify the definitions used and category choices offered.
• How is this information collected? Self-reported? Reported by clinician or other third party?
• At what level of detail (i.e., individual patient discharge, aggregate) does your agency require this information to be reported?
• What organizations (i.e., hospitals, health plans, physician organizations) report this information to your agency?
• What race and ethnicity data are voluntary? Required?
• What is the approximate level of compliance with these reporting requirements?
• Are you aware of differences in race and ethnicity data currently collected by your agency and others within your state? If so, please specify.
• How are the data your agency collects currently used in your state? To whom are they reported?
• What other entities collect race and ethnicity data in your state, if known? Do you currently cooperate with any of these entities for data collection? Reporting? Other purposes?
• Who has the authority to make a decision about the data that your agency collects? Are data elements governed by law, regulation, agency authority, other?
• What stakeholders would be opposed to changes to data your agency currently collects?
• How would changes in data collection requirements affect implementation of and compliance with data collection? Within your agency?

Within hospitals, health plans, and physician organizations reporting data to your agency?

- How would changes affect your reporting of data currently collected?
- How long do you expect it would take to agree on data standards across agencies in your state? What would require the most time? What is the greatest unknown?
- How long would it take your agency to implement the data requirements? What would require the most time?
- Can you quantify or estimate the costs to your agency and the organizations reporting to your agency of any of the changes necessary to implement new data collection requirements?

Benefits: What would be the benefits of enhancing and standardizing race and ethnicity data collection in your state? Nationally?

- What is the intended purpose(s) of the race and ethnicity data your agency currently collects?
- What qualitative benefits to your agency, the organizations reporting to your agency, other government entities, beneficiaries of public programs, or the public (i.e., stakeholders) do you foresee as a result of more detailed or standardized data collection? Better service provision? Greater comparability in reporting? Opportunities for improved quality of care?
- Can you identify any quantitative benefits to your agency or other stakeholders that might result from more detailed or standardized data collection? Less supplementary data collection? Opportunities for economies of scale? Simplification? Improved ability to target public or private programs?
- To whom would these benefits accrue? Your agency? Other state or local agencies? Hospitals and other health care providers? Public programs or program beneficiaries?
- How long do you expect it would take for your agency or other stakeholders to achieve these benefits? Immediate? Less than one year? Less than five years? Longer than five years?
- Can you quantify or estimate the savings or other benefits to your agency or other stakeholders due to more detailed or standardized data collection?

Thank you for your input.

Appendix F

Collection of Data on Race and Ethnicity by Private-Sector Organizations: Hospitals, Health Plans, and Medical Groups

*David Nerenz and Connie Currier**

BACKGROUND

The published literature on racial and ethnic disparities in health, access to health care, and quality of care paints a distressing picture of inequities. Members of minority groups are generally less likely than their non-Hispanic white counterparts to have health insurance coverage, a regular physician, good control of chronic illnesses, access to invasive diagnostic or surgical procedures, adequate pain medications, or appropriate therapy for mental health problems (IOM, 2003). The higher prevalence of chronic conditions like hypertension and diabetes in some minority groups, coupled with problems in access to care for conditions like cancer, heart disease, or stroke, leads to significant disparities in measures of mortality and life expectancy (Satcher, 2001).

Although some of the disparities in broad measures of population health—e.g., life expectancy—may be related to socioeconomic factors such as income, education, and access to health insurance, there is ample evidence of disparities in care provided to individuals with similar jobs, similar health insurance coverage, and similar incomes (Fiscella et al., 2000). There is evidence, for example, of racial and ethnic disparities in the stage of breast cancer at diagnosis (presumed to at least partially reflect screening

---

*David R. Nerenz, Ph.D., and Connie Currier, Dr PH., are project director and assistant professor, respectively, at the Michigan State University Institute for Health Care Studies.

services) and survival among members of the same managed care plan (Yood, Johnson, and Blount, 1999).

When individuals who live in the same general area and share many socioeconomic characteristics other than race or ethnicity receive different patterns of care, explanations for these differences are likely to be found in the dynamics of physician decision making, doctor-patient interaction, individual attitudes and beliefs about illness and treatment, or organizational characteristics of health care systems (Grantmakers in Health, 2000).

There have been some studies documenting differences in physician recommendations or referral patterns for cardiac surgery as a function of patient race or gender, even when other characteristics were controlled (Grantmakers in Health, 2002). Studies of disparities in dose of adjuvant chemotherapy for women with breast cancer have also identified individual physician decision making as a key factor (LaVeist, Morgan, and Arthur, 2002). When the causal factors related to disparities cannot be identified so closely with specific individuals, though, there are a number of characteristics of organizations or health care systems that have been linked to either the current existence of disparities or their potential reduction or elimination. These factors include institutional racism (Williams and Rucker, 2000) cultural competence (Cohen and Goode, 1999), or "patient-centeredness (Picker Institute)."

As research and health policy attention shifts from the documentation of disparities to the testing of initiatives to reduce disparities, the role of organizations such as hospitals, health plans, and medical groups will be crucial. Each currently plays an important role in quality measurement, quality improvement, and establishing the norms, values, and systems for accountability of medical care. If we follow the recommendation of a recent IOM panel and view disparities as an important quality of care problem (IOM, 2003), then it is important to understand how hospitals, health plans, and medical groups can play a role in reducing disparities.

In the domains of quality measurement and quality improvement, these organizations already play important roles, as indicated below.

Hospitals:
- Responsible for assessing and ensuring quality for inpatient care
- Able to establish and enforce clinical policies and guidelines
- Able to produce clinician profiles for quality and cost of care
- Able to grant or rescind privileges
- Subject to external review for both accreditation and licensing purposes

Health plans:
- Accountable to public and private purchasers for achieving stan-

dards of quality in terms of both processes and outcomes of care
  • Subject to formal accreditation processes that include use of standard
quality measures Health Plan Employer Data and Information Set (HEDIS)
and member survey data Consumer Assessment of Health Plans (CAHPS)
  • Responsible for implementing clinical guidelines and designing disease management programs
  • Able to use a variety of financial and other incentive systems to alter
clinician behavior

Medical Groups:
  • Able to enforce clinical guidelines and policies
  • Able to select group members on the basis of compatible practice
styles
  • Able to maintain medical records systems, registries, and other sorts
of data systems used to measure quality of care
  • Able to design and implement compensation systems that reward
quality

  From the perspective of the purchaser, responsibility for assessing and
improving quality of care has largely been delegated to health plans and
hospitals. For health plans, requirements for National Committee for Quality Assurance accreditation and comparison using HEDIS measures and
CAHPS results represent methods by which purchasers attempt to improve
quality of care. For hospitals, more recent initiatives by the Leapfrog Group
and other purchasers have focused on technologies such as electronic order
entry for pharmacies or the "volume-outcome relationship" to improve
quality in inpatient settings. Joint Commission on Accreditation of Healthcare Organizations (JCAHO) accreditation requires that a "cultural assessment" be conducted in the context of patient and family education to
ensure that social, ethnic, cultural, and emotional factors are considered in
providing patient care (JCAHO, 1999). Medical groups have not been as
directly engaged as agents for quality improvement, except perhaps in
examples like the Buyers Health Care Action Group in the Minneapolis/
St. Paul area, where a "direct contracting" model has been used to try to
manage both cost and quality (Buyers Health Care Action Group, 2002).
  When these types of organizations engage in quality improvement activities, it is generally presumed that they have the necessary clinical, administrative, and patient demographic information to create effective programs and monitor outcomes. Health plans working to improve breast
cancer screening rates, for example, are presumed to have information that
will enable them to identify women over the age of 50 who either have or
have not had a mammogram during a certain time period. Hospitals working to improve use of beta-blockers after acute myocardial infarction are

expected to be able to identify patients with a diagnosis of acute MI and to know whether or not beta-blockers were prescribed at discharge.

If health plans, hospitals, or medical groups are going to be engaged as active partners in reducing racial and ethnic disparities in care, it is important to know whether they have the essential data on the race or ethnicity of members or patients in order to: (a) assess disparities, (b) identify individuals in target populations for intervention, (c) evaluate the effects of those interventions, and (d) be accountable for reducing or eliminating disparities.

The availability of data on race, ethnicity, and primary language is crucial to these organizations' ability to focus their efforts on quality improvement and reduction or elimination of disparities. Initiatives that seek to expand translation services, community outreach programs, access to care, and quality improvement will be much more beneficial to members of racial, ethnic, or linguistic minority groups if they are guided by information on the groups most affected by disparities and the extent to which specific structural or process variables are responsible for those disparities.

From the larger perspective, standardized data collection and reporting on patient race and ethnicity are essential to identify discriminatory practices in the process, structure, and outcomes of care, to demonstrate compliance with civil rights legislation, and to measure achievement and monitor progress toward DHHS Healthy People 2010 goals (IOM, 2003).

In this paper, we will attempt to summarize what is known about the extent to which hospitals, health plans, and medical groups collect data on race/ethnicity and how they use that information to either assess or attempt to eliminate disparities in care. When organizations do collect and/or use information, we will attempt to describe the accuracy and completeness of that information and the methods used to obtain it.

HOSPITALS

In February 2003, a survey was sent to a nationally representative sample of 1,000 hospitals to determine whether they collected data on race and ethnicity, and the factors that went into the decision to collect the data. A total of 262 completed surveys were returned. Seventy-nine percent of respondents reported that they do collect racial and ethnic data about patients. The patient was identified as the primary source of information about race or ethnicity, although a large percentage of respondents reported that the admitting clerk obtained the information by observation. Information was most often collected upon admission. A complete summary of the results of the survey can be found in the survey summary at the end of this appendix.

Hospitals have been required to provide care without regard to the race or ethnicity of patients since 1966 with the implementation of regulations based on the 1964 Civil Rights Act (Smith, 1999). Title VI of the act required hospitals to be able to document absence of discriminatory treatment through the submission of an annual survey form (Assurance Form 441) and possible site visits. An Office of Equal Health Opportunity (OEHO) was established in the surgeon general's office to monitor compliance. The requirement was imposed because there were patterns of blatant segregation in both northern and southern hospitals until then, evidenced in either fully segregated hospitals or segregated units and programs within hospitals.

The passage of the Civil Rights Act and subsequent enforcement actions by the OEHO (or credible threats of such actions) served to eliminate the most overt forms of segregation and discrimination in relatively few years, primarily because the subsequent passage of Medicare and Medicaid legislation gave federal officials a powerful financial tool to use to force compliance with Title VI requirements. Hospitals were threatened with ineligibility for Medicare and Medicaid reimbursement if they failed to comply, and even recalcitrant hospitals in the Deep South eventually gave in (Smith, 1999).

The requirement to document nondiscriminatory practices did not apparently include a specific requirement to collect data on the race or ethnicity of individual patients. Gross violations of Title VI in the mid-60, such as the presence of racially segregated units or separate dining rooms, did not require the presence of data on individual patients in order to be apparent. In the late 1970s, the Office for Civil Rights and Health Care Financing Administration were still working on the development of hospital surveys that would include information on the race and ethnicity of patients in order to monitor compliance with Hill-Burton Act requirements.[1] In the mid-1990s, the uniform hospital claim forms (UB-92 and HCFA-1500) did not include a data field for race or ethnicity in spite of strong efforts by advocacy groups to have this information included. HCFA argued at the time that information on race or ethnicity could be obtained from Social Security enrollment files (ibid). It appears, then, that although many hospitals developed mechanisms for collecting data on the race or ethnicity of patients as a way to document compliance with Title VI, such data collection has not been universally implemented or enforced.

---

[1]The Hill Burton Act, or Hospital Survey and Construction Act, passed in 1946, originally allowed the construction of racially separated hospitals if the planned facility, equitably provided, on the basis of need, facilities, and services of like equality. The provision was later challenged in *Simkins v. Moses Cone Memorial Hospital* and overridden by Title VII of the Civil Rights Act.

There appears to be considerable variation from state to state and region to region in interpretation of current requirements and practices. An attorney at a large hospital in Michigan who had previously worked in the DHHS Office for Civil Rights (OCR) stated that hospitals were technically required to collect data, but that there was little or no incentive for doing so and that there was no active enforcement of the requirement.[2] An official at another large hospital in Michigan stated that the hospital did not routinely collect data on race or ethnicity and that he was not aware of a legal requirement.

Even if enforcement of a legal requirement is minimal, it appears that large numbers of hospitals do collect data on race or ethnicity and incorporate that information into patient registration systems and discharge abstracts. According to an official at the Michigan Health and Hospital Association, the Michigan Inpatient Database (a collection of discharge abstracts from essentially all general medical/surgical hospitals in the state) includes racial and ethnic data for approximately 75 percent of patient records (Robert Zorn, 2002, personal communication, Sparrow Regional Health System).

A study conducted in 1993 in New York found data on race or ethnicity assigned to virtually all patients whose records were reviewed in a study on care for acute myocardial infarction (Blustein, 1994). The accuracy of the data was called into question, though, when records were compared for patients who had two separate admissions in the same study period. Although there was agreement in classification for over 93 percent of the patients with two admissions, kappa statistics for agreement were relatively low for groups other than black or white. The authors of the study observed that the designation of race or ethnicity was generally assigned by an admitting clerk or some other administrative staff person based on his or her own judgment rather than on the basis of patient self-report, and cited a prior study from California making the same observation (California Department of Health, 1990).

The Hospital Cost and Utilization Project (HCUP) databases, which have been used for several studies of racial and ethnic disparities in use of procedures (Harris, Andrews, and Elixhauser, 1997), are created from discharge abstract data from hospitals in approximately 30 states. Patient's race is one of the fields in the HCUP National Inpatient Sample (NIS) database, suggesting strongly that race is an element of the discharge abstract or underlying medical record in most hospitals (at least those hospitals in states that maintain state discharge abstract systems and were included in HCUP-NIS). Race is also an element included by 12 of 18 states

---

[2]Personal communication, Peter Jacobson, J.D, M.P.H., November 15, 2002.

included in the State Inpatient Database (SID) in HCUP. Studies of hospital care using HCUP can identify racial and ethnic disparities, but some states cannot be included in the analyses. In one study of disparities in care for patients with AIDS using HCUP, for example, the investigators had to delete Illinois from the analysis because no data on the race or ethnicity of patients were available from Illinois (Hellinger and Fleishman, 2001).

Analyses of Medicare claims files have been conducted frequently to identify racial and ethnic disparities in care (McBean and Gornick, 1994). However, the racial and ethnic data in these files should be used with caution. Data on race and ethnicity for these Medicare Provider and Review (MEDPAR) files come from the Medicare Enrollment Database (EDB), populated from the Social Security Administration (SSA) Master Beneficiary Record (MBR), which carries only four racial codes: white, black, other, and unknown. The Medicare EDB is updated using information from another SSA file called the NUMIDENT file that carries seven codes: white, black, other, unknown, Asian, Hispanic, Northern American Indian or Alaska Native. The racial and ethnic categories of the Medicare EDB are updated periodically using data from the SSA NUMIDENT file, and also by mailings to certain ethnic-sounding names once a year, though not on a set schedule.[3] Procedures for updating the EDB do not guarantee that all records get updated on a regular basis. Many minority beneficiaries remain misclassified in the Medicare EDB (Arday et al., 2000).

Even if some hospitals do not collect routinely data on race or ethnicity for all patients, virtually all hospitals collect such data for special purposes such as birth and death certificates, tumor registries, and reportable diseases like HIV/AIDS. A study of the completeness and validity of information on AIDS cases provided by hospitals to the CDC, for example, found 83 percent concordance between information on race or ethnicity in the original case report and information found subsequently in a medical record review (Klevens, Fleming, and Lee, 2001). An analysis of the accuracy of racial or ethnic categorizations in California hospital birth certificate data found generally complete and valid information when compared against subsequent self-report in the context of a postpartum follow-up survey of mothers, but also noted that birth clerks used a mix of mothers' self-identification and their own observations as methods for assigning race and ethnicity (Baumeister, Marchi, and Pearl, 2000).

Some of these studies are now 10-15 years old, and procedures for the collection of data on race and ethnicity may have changed in the interim because of changes in categories used in the U.S. Census, attention to the

---

[3]Personal communication, Medicare information from Medicarestats@cms.hhs.gov, December 18, 2002.

issue of race and ethnicity in the 1998 DHHS initiative on disparities,[4] or attention to the collection and usage of personal information in the context of HIPAA (IOM, 2003).

HEALTH PLANS

In March 2003, the survey was sent to a sample of 158 health plans to better understand the current state of data collection related to race and ethnicity; a total of 38 completed surveys were returned. Although we are reluctant to draw conclusions from such a relatively small sample, it appears that health plans are less likely than hospitals to collect data on enrollees' race and ethnicity. Those that do, use the information primarily to translate materials, for quality improvement purposes, and to inform disease management programs. Health plans that do not collect data expressed concern about enrollees' perceptions regarding the need for and use of such data. Both health plans that do and do not collect data are concerned about the lack of a reliable system for data collection.[5] A complete summary of the survey results can be found in at the end of this appendix.

None of the various types of health plans—health maintenance organizations, preferred provider organizations, indemnity plans, or point-of-service plans—have been required to collect data on the race or ethnicity of their members as part of any federal legislation. In fact, most of the legal issues related to insurance companies' collection and use of data on race and ethnicity have been based on concerns about how such information might be used to adversely affect services offered to members of minority groups. Concerns about insurance "redlining" and charging higher premiums to members of minority groups have led several states to either pass laws or develop language in state insurance regulations limiting or prohibiting companies' ability to ask questions about race or ethnicity.

These concerns, which typically arose in insurance areas other than health insurance, led many health plans surveyed in the context of a 1999 DHHS conference on health plans' use of data on race and ethnicity to express the belief that it was illegal for them to collect such data (Bierman et al., 2002). Subsequent reviews of both federal law (Perot and Youdelman, 2001) and applicable state laws (Berry et al., 2001) have shown that there are no federal laws or regulations barring health plans from collecting data on the race or ethnicity of their members, and that only four states have

---

[4]U.S. Department of Health and Human Services. *The Initiative to Eliminate Racial and Ethnic Disparities in Health.* http://raceandhealth.hhs.gov. Accessed October 15, 2002.

[5]Personal communication, Deborah Wheeler, deputy director, Quality Initiatives and Industry Standards and Teresa Chovan, director, Policy Research, American Association of Health Plans, 04/23/03.

laws that can be interpreted as prohibiting such data collection. In those four states, the specific language of the laws typically addresses the collection of data on race or ethnicity at the time of application for insurance, and does not address health plans' ability to collect data at a later time (for example, in the context of a health risk appraisal for new members).

An informal survey of health plan representatives invited to attend the 1999 DHHS meeting, and a concurrent telephone and e-mail survey of large not-for-profit health plans involved in the HMO Research Network, indicated that essentially no plans routinely collected data on the race or ethnicity of their members at that time. Some plans had done special research projects or Quality Improvement initiatives that had involved collection of that information, but no plans reported the routine collection of racial or ethnic information as part of the regular membership process. In addition to the perceived legal barriers, plans cited concerns about potential adverse reactions in minority communities, or the risk of adverse publicity associated with finding disparities in care, as reasons for not collecting the information. Virtually all of these plans were organized as not-for-profit HMOs, but there is no reason to believe that the policies and practices of for-profit HMOs, PPOs, or other forms of managed care are fundamentally different.

Subsequent discussions with the National Committee for Quality Assurance (NCQA) indicated that, because of the lack of consistent data on race and ethnicity available to managed-care plans, NCQA did not require collection of these data as part of its accreditation process. Later communications with Dr. Greg Pawlson of NCQA indicated that the committee is exploring a variety of approaches to increasing the reporting of data on race and ethnicity in the context of quality assessment.

A recent demonstration project conducted with 15 managed-care plans from different parts of the country showed that the health plans were able to obtain data on the race or ethnicity of their members through a variety of mechanisms, although none had data routinely available at the start of the project. Plans were able to obtain these data from state Medicaid program enrollment files, from medical records and patient registration systems of contracting medical groups, from self-report items in member surveys, and through the use of surname-recognition software that could distinguish Hispanic from non-Hispanic plan members (Nerenz, Gunter, and Garcia, 2002). Plans were then able to link the information to procedures for producing HEDIS and CAHPS reports in order to stratify those reports by race and ethnicity.

MEDICAL GROUPS

In early March 2003, a survey was sent to 250 medical groups to gather information about current practices and policies regarding the collection of data on race and ethnicity; a total of 83 valid surveys were returned. Like health plans, medical groups are less likely than hospitals to collect data on patients' race and ethnicity. Those groups that do collect data do so primarily for internal quality improvement or disease management purposes rather than because of an external reporting requirement. Groups that do not collect data believe that it is either unnecessary or potentially troubling to patients, or that it cannot be done reliably. A complete summary of survey results can be found at the end of this appendix.

In the 1964 Civil Rights Act and subsequent implementing regulations (Smith, 1999), physician offices and group practices were exempted from any requirements to collect data on patient race or ethnicity or to document nondiscriminatory practices. Therefore very little is known about the current policies and practices of medical groups.

Some multispecialty group practices that have been closely affiliated with hospitals have developed procedures for collecting data on race and ethnicity as part of the patient registration process. Those data have been used for studies of racial and ethnic disparities in patterns of care and disease outcomes. For example, the Henry Ford Medical Group in Detroit has been able to use data on race and ethnicity in its administrative databases to support studies of breast cancer survival rates and patterns and outcomes of care for patients with diabetes.

CONCLUSIONS

Collection of data on race and ethnicity seems to be far more common, routine, and complete in hospitals than in health plans or medical groups. We presume that the main reason for this difference is the legal requirements imposed on hospitals by the 1964 Civil Rights Act, even if the enforcement of those requirements is not very stringent today and even if the original requirements did not specifically include collection of data on the race or ethnicity of individual patients. State-level requirements for collection and reporting of data seem to be an important driving force for those hospitals that do collect racial and ethnic data.

The regulations associated with the Health Insurance Portability and Accountability Act (HIPAA) will have an effect on the collection and use of demographic information, including data on race and ethnicity. It is not clear, though, whether HIPAA will serve to promote or inhibit the collection and use of such data overall. On one hand, HIPAA has made health care organizations extremely cautious about their policies for collection, communication, and use of all forms of personally identifiable health information. This atmosphere of caution will probably inhibit organizations that might otherwise move toward collection and use of data to reduce racial, ethnic, or SES disparities in care. On the other hand, the protections built into HIPAA may have the effect of reassuring individuals and community groups that information provided will be used for legitimate health care and quality improvement purposes and will not be used inappropriately. Until some years of experience with HIPAA regulations have accumulated, though, it will be difficult to say with any certainty how these competing forces will balance out.

Key questions remain about the procedures used to collect and categorize information on race and ethnicity, and about the extent to which practices in hospitals or other organizations conform to current U.S. Census Bureau standards and the recommendations of expert panels and advisory groups. Specifically, it will be important to determine:

- the extent to which categories and procedures used to define race and ethnicity conform to categories and procedures used in the 2000 U.S. census or to the recommendations made in OMB Directive #15 (Ford et al., 2002; Friedman et al., 2000);
- the extent to which local variations from those categories and procedures (e.g., "fine-grained" categorization of Hispanic or Asian groups) are either used at all or used in ways that would allow "roll-up" to broader U.S. census categories;
- the extent to which racial and ethnic information is obtained on the basis of patient or health plan member self-report (preferred) versus on some other basis such as visual observation on the part of a registration clerk;
- the extent to which information on race and ethnicity is complete and accurate as assessed by agreement with other data sources; and
- the extent to which information on race and ethnicity is used in the context of quality improvement and/or disparity reduction initiatives.

These issues of "process" for the collection of data on race and ethnicity (and other demographic factors such as SES) are extremely important if health care organizations are going to work together in any sort of collaborative fashion on disparities, if research is going to be conducted on disparity reduction initiatives, and if governmental agencies and accrediting bodies are going to effectively monitor and encourage those initiatives. A single organization could conceivably use any method it wants to collect, categorize, and use demographic information, as long as it does so in a way that fits local circumstances and is consistent over time. Comparison of data across organizations, though, requires broad adherence to standards for methods of collecting and categorizing data. It is possible to have standardization *and* flexibility to adapt to local circumstances (e.g., multiple subgroups of Hispanic or Asian patients in a particular community) by using methods to "roll up" fine-grained categories into more global categories for aggregate analysis (Ford et al., 2002).

Patients and health plan members will have legitimate concerns about why they are being asked to provide information about their race or ethnicity and other demographic factors, and how this information will be used. In the absence of a long-standing and visible tradition of use of this information to improve quality of care for minority groups, it will be reasonable for individuals and community groups to fear some form of discriminatory treatment. Over time, the most effective way to address this concern will probably be to show in very concrete ways how the information is being used to improve and expand, rather than deny, services to members of minority groups. In the near term, though, several general strategies for reducing concern can be identified:

- Involve community leaders in all aspects of the planning and design of processes for data collection, data analysis, and use of data for quality improvement.
- Take all possible opportunities to communicate with community groups about the reasons for collection of data and the use of data for quality improvement purposes.
- Collect the data in a meaningful context—for example, a new patient information form or a new plan member health risk appraisal survey.
- Don't break promises—if data on race, ethnicity, or SES are intended to be used for a specific purpose such as expansion of translation services, then make sure those service expansions actually occur and are visible in the community.

SUMMARY

Disparities among racial and ethnic groups on measures of health, access to health care, and quality of care have been well documented. Many underlying reasons for disparities have been identified, but systematic efforts to reduce or eliminate disparities are relatively new, so not much is known about their effectiveness. Initiatives to reduce disparities that involve private-sector health insurers or delivery organizations generally must often rely on those organizations' collection of data on race and ethnicity to support either key features of program design (e.g., identifying members of target populations for intervention) or program evaluation. Hospitals have been required to document nondiscriminatory treatment of patients since the mid-60s, but procedures for the collection of racial and ethnic data on patients in support of that documentation requirement vary from hospital to hospital, and enforcement of policies on data collection can be weak to nonexistent. Health plans and medical groups are not legally required to collect racial or ethnic data, some states prohibit data collection, and most have no formal process for doing so. Data collected for special purposes, though (e.g., CAHPS data for health plans; disease registries for medical groups) can be used effectively for initiatives aimed at reducing disparities.

ACKNOWLEDGMENTS

Several surveys were conducted in preparation for this paper. These surveys obtained information on data collection practices of hospitals, health insurance plans, and medical group practices. Romana Hasnain-Wynia, Ph.D., of the Health Research Education Trust coauthored the survey of American Hospital Association members. Julia Sanderson-Autin, R.N., of the American Medical Group Association coauthored the survey of members of that organization.

REFERENCES

Arday, S.L., D.R. Arday, S. Monroe, and J. Zhang
   2000   HCFA's racial and ethnic data: Current accuracy and recent improvements. *Health Care Financing Review* 21:107-116.
Baumeister, L., K. Marchi, and M. Pearl
   2000   The validity of information on "race" and "Hispanic ethnicity" in California birth certificate data. *Health Services Research* 35:869-883.
Berry, E., S. Hitove, J. Perkins, D. Wong, and V. Woo
   2001   *Assessment of State Laws, Regulations, and Practices Affecting the Collection and Reporting of Racial and Ethnic Data by Health Insurers and Managed Care Plans.* Presented at the Annual Meeting of the American Association of Health Plans, Washington, DC.
Bierman, A.S., N. Lurie, K. Scott Collins, and J.M. Eisenberg
   2002   Addressing racial and ethnic barriers to effective health care: The need for better data. *Health Affairs* May/June(3):91-102
Blustein, J.
   1994   The reliability of racial classifications in hospital discharge abstract data. *American Journal of Public Health* 84:1018-1021.
Buyers Health Care Action Group
   2002   A health care trailblazer says *Harvard Business Review. Buyers Health Care Action Group Newsletter* 1(2):September.
California Department of Health
   1990   *Report of Results of the OSHPD Reabstracting Project.* Sacramento, CA: California Department of Health, Office of Statewide Planning and Development.
Cohen, E., and T.D. Goode
   1999   *Policy Brief 1: Rationale for Cultural Competence in Primary Health Care.* Washington, DC: National Center for Cultural Competence, Georgetown University Child Development Center.
Fiscella, K., P. Franks, M.R. Gold, and C.M. Clancy
   2000   Inequality in quality: Addressing socioeconomic, racial, and ethnic disparities in health care. *Journal of the American Medical Association* 283:2579-2584.

Ford, M.E., D.D. Hill, D. R. Nerenz, M. Hornbrook, J. Zapka, R. Meenan, S. Greene, and C.C. Johnson
    2002   Categorizing race and ethnicity in the HMO Cancer Research Network. *Ethnicity and Disease* 12:135-140.
Friedman, D.J., B.B. Cohen, A.R. Averbach, and J.M. Norton
    2000   Race/ethnicity and OMB Directive 15: Implications for state public health practice. *American Journal of Public Health* 90:1714-1719.
Grantmakers in Health
    2002   *Racia/lEthnic Differences in Cardiac Care: The Weight of the Evidence.* Washington, DC: Henry J. Kaiser Family Foundation.
    2000   *Strategies for Reducing Racial and Ethnic Disparities in Health.* Washington, DC: Grantmakers in Health.
Harris, D.R., R. Andrews, and A. Elixhauser
    1997   Racial and gender differences in use of procedures for black and white hospitalized adults. *Ethnicity & Disease* 7:91-105.
Hellinger, F.J., and J.A. Fleishman
    2001   Location, race, and hospital care for AIDS patients: An analysis of 10 states. *Inquiry* 38:319-330.
Institute of Medicine
    2003   *Unequal Treatment: Confronting Racial and Ethnic Disparities in Health Care.* B.D. Smedley, A.Y. Stith, and A.R. Nelson, editors. Committee on Understanding and Eliminating Racial and Ethnic Disparities in Health Care. Board on Health Sciences Policy. Washington, DC: The National Academies Press.
Joint Commission on Accreditation of Healthcare Organizations
    1999   *Using Performance Measurement to Improve Outcomes in Behavioral Health Care.* Oakbrook Terrace, IL: Joint Commission on Accreditation of Healthcare Organizations.
Klevens, R.M., P.L. Fleming, and J. Li
    2001   The completeness, validity, and timeliness of AIDS surveillance data. *Annals of Epidemiology* 11:443-449.
LaVeist, T.A., A. Morgan, and M. Arthur
    2002   Physician referral patterns and race differences in receipt of coronary angiography. *Health Services Research* 37:949-962.
McBean, A.M., and M. Gornick
    1994   Differences by race in the rates of procedures performed in hospitals for Medicare beneficiaries. *Health Care Financing Review* 15:77-85.
Nerenz, D.R., M. Gunter, and M. Garcia
    2002   *Quality of Care for Underserved Populations Developing a Health Plan Report Card on Quality of Care for Minority Populations.* New York: The Commonwealth Fund.
Perot, R.T., and M. Youdelman
    2001   *Racial, Ethnic, and Primary Language Data Collection in the Health Care System: An Assessment of Federal Policies and Practices.* New York: The Commonwealth Fund.
Satcher, D.
    2001   Our commitment to eliminate racial and ethnic health disparities. *Yale Journal of Health Policy, Law, and Ethics* (1):1-14.
Smith, D.B.
    1999   *Health Care Divided: Race and Healing a Nation.* Ann Arbor, MI: University of Michigan Press.

Williams, D.R., and T.D. Rucker
   2000   Understanding and addressing racial disparities in health care. *Health Care Financing Review* 21:75-90
Yood, M.U., C.C. Johnson, and A. Blount
   1999   Race and differences in breast cancer survival in a managed care population. *Journal of the National Cancer Institute* 91:1487-1491.

SUMMARY OF SURVEY OF
AMERICAN HOSPITAL ASSOCIATION HOSPITALS

In 2003, the Institute of Medicine released the report *Unequal Treatment: Confronting Racial and Ethnic Disparities in Health Care*, which highlighted the importance of collecting patient data by race, ethnicity, and primary language. The report noted that if such data were collected, researchers, policy makers, and clinicians could disentangle factors associated with health care disparities, facilitate monitoring of performance, ensure accountability to health plan members and health care purchasers, improve patient choice, and identify discriminatory practices. The growing body of evidence documenting disparities in health and health care underscores the importance of such data collection. Fortunately, a number of public and private organizations, including the American Association of Heath Plans, are beginning to consider how to incorporate minority data collection into overall quality measurement initiatives.

Currently, information about racial and ethnic data collection in hospitals is very limited. We know that overall the reasons for disparities are poorly understood and that there are significant gaps in our understanding. One critical problem in understanding these reasons is the lack of uniform data by race, ethnicity, and primary language at the hospital and health system level. Currently, the methods for data collection by race, ethnicity, and primary language are inconsistent and incompatible across most hospitals and health systems.

The Health Research and Educational Trust (HRET), an affiliate of the American Hospital Association, has a track record of working with hospitals in diverse communities across the country. HRET is currently leading a research project with support from the Commonwealth Fund, to develop a Uniform Framework for Collecting Race, Ethnicity, and Primary Language Data in Hospitals. As part of this project, HRET is working closely with a consortium of six hospitals and health systems.

In an effort to better understand the current state of data collection in hospitals and to inform the National Academies Panel on DHHS Collection of Race and Ethnicity Data, in February 2003, a survey on the collection of such data was developed collaboratively by the Michigan State University Institute for Health Care Studies and the Health Research and Educational Trust. The survey was sent to a nationally representative sample of 1,000 hospitals and was designed to determine whether hospitals collected data on race and ethnicity and what factors go into the decision to collect the data. A total of 262 completed surveys have been returned to date, and we provide a preliminary analysis based on these responses.

Survey Content

Respondents were asked to indicate *who provides information* about patients' race and ethnicity and *when it is collected* for up to three units or clinics that they identified within the organization. After reviewing the units listed by respondents, four categories of units were created: (1) admitting/ registration, (2) emergency department, (3) outpatient/specialty clinics, and (4) hospital (general).

Respondents were asked to "check all that apply" for the question, "Who provides information about the patient's race or ethnicity?" using the following categories:

- Patient self-identifies
- Caretaker/guardian provides information
- Admitting clerk obtains information from patient
- Admitting clerk provides information based on observation
- Health care provider obtains information from patient
- Health care provider provides information based on observation
- Don't know

Respondents were asked, "When is the information collected?" and again to "check all that apply" using the following categories:

- Upon admission
- At discharge
- At first visit/new patient registration
- Included in health care provider's discharge notes/medical record
- Don't know

Findings

The majority of hospitals (79 percent) reported that they collect racial and ethnic data about patients. Fifty-seven percent of those respondents indicated that more than one unit or clinic within the hospital collected data.

The *patient* was identified as the primary source of information on race or ethnicity for all categories of units. Either an admitting clerk obtains the information from the patient and completes a form or inputs the information into a computer (67 percent of responses among hospitals collecting data at all, $n = 206$), or the patient self-identifies by completing a form him or herself (65 percent). In the emergency department and admitting/registration, the information was more often obtained by an admitting clerk. In outpatient/specialty clinics and hospitals (general), the patient was more likely to self-identify.

Across all units, 53 percent of respondents reported that information was provided by a *caretaker or guardian*. Information was more likely to be obtained from a caretaker or guardian in a hospital (general) or emergency department than in other units. The admitting clerk frequently obtained information by observation of the patient's race or ethnicity (51 percent), though this occurred most often in the emergency department.

Information was most often collected *upon admission* for all units (checked 85 percent of the time), or *at first visit/new registration* (59 percent). It was collected less often through health care provider notes or the medical record (11 percent), and rarely at discharge (2 percent). These percentages were fairly consistent across all units. Racial and ethnic information is also collected during the hospital stay, upon subscribing to an HMO, during preadmission screening, at an initial intake assessment, when obtaining a birth certificate, and when making an initial appointment. A referring facility may also provide data on a patient's race and ethnicity.

Hospitals were asked *why they established policies/practices* to collect data on patient race and ethnicity and were given the option to check all responses that applied. The largest single set of respondents reported that their hospital established policies to collect data on patient race and ethnicity because it was *required by law or regulation* (42 percent). Twenty-three percent reported collecting the data for quality improvement, 19 percent felt it was important for community relations, and 12 percent reported collecting it because it helped target marketing efforts. Seventeen percent indicated other reasons for collecting these data such as: it is required by the state; it is included as basic demographic information for medical records or a hospital database or cancer registry; it is required by the state hospital association; and it is used for teaching and for conducting research on best practices, trends, and preventive care.

Eighty-six percent of respondents indicated that they provide specific categories for patients or guardians to check off when data on race and ethnicity are collected, and 14 percent of the respondents reported collecting the information using a "fill in the blank" open question. For hospitals that reported using specific categories, respondents were asked which categories they used from the minimum racial and ethnic classifications used by the Census Bureau (American Indian and Alaska Native, Asian, Black or African American, Native Hawaiian and Other Pacific Islander, and White), and other racial and ethnic categories that were reported in the profile of general U.S. demographic characteristics from the 2000 census. Again, respondents were asked to "check all that apply." Respondents were also given the option to indicate any additional categories they used that were not included among those used on the U.S. census. Respondents added Cambodian, Czech, Hindu, Hmong, Laotian, Middle Eastern, Persian, Polish, Portuguese, Russian, Thai, and Ukrainian.

None of the broader categories used to specify race or ethnicity—i.e., white, black, Hispanic, Asian, Native American and Alaska Native—were used by more than 95 percent of the respondents, and a number of the smaller, more "fine-grained" categories were used by 4 percent or more of the respondents. These two observations suggest that hospitals do some significant tailoring of the standard U.S. census categories to adjust to local circumstances.

Specific racial and ethnic categories and the percentage of hospitals that used them are listed below.

Caucasian/White	95%
African American/Black	94%
Spanish/Hispanic/Latino	81%
American Indian	77%
Asian	77%
Alaska Native	26%
Other Pacific Islander	24%
Mexican, Mexican American, Chicano	11%
Native Hawaiian	9%
Asian Indian	8%
Chinese	8%
Filipino	7%
Japanese	7%
Puerto Rican	5%
Cuban	4%
Vietnamese	4%
Samoan	4%
Korean	4%
Guamanian or Chamorro	1%

Hospitals were asked in what percentage of cases data on race and ethnicity were *missing or unavailable*. Respondents gave the widest possible range of responses, from 0 to 100 percent. The 100 percent figure presumably represents data that are not collected in this area at all, thus indicating 100 percent missing; the "0 percent" figure seems too good to be true, if in fact this indicates that all data fields for race and ethnicity are always filled. Another more likely interpretation could be that in that hospital there is at least some entry in the data field for race or ethnicity for every patient, regardless of whether the entry is "unknown" or "other."

Eighty-nine percent of the respondents reported that race and ethnicity data are stored in an *electronic database* in their hospital. Hospitals were asked who may obtain access to the data and how the hospital uses the data that are collected. Again, a number of possible answers were provided, and respondents were asked to check all that applied. Seventy-nine percent of respondents reported that hospital employees had access to the data, followed by health care providers (41 percent), researchers (15 percent), grantees/contractors (8 percent), and the public 3 percent. Five percent of respondents said they did not know who had access to the data.

Hospitals were asked *how data on race and ethnicity are used*. Responses indicated that they used for a variety of internal purposes, including ensuring the availability of interpreter services (36 percent), quality improvement or disease management programs (36 percent), program/benefit design (17 percent), marketing (13 percent), actuarial purposes (2 percent), and underwriting (1 percent). Data are also shared with federal, state, and local governmental agencies including state health departments (36 percent), Medicare (27 percent), Medicaid (26 percent), local health departments (16 percent), and the Veterans Administration (10 percent). Nongovernmental agencies/organizations are also given access to the data, including accrediting bodies (21 percent), community groups (5 percent), and purchasers (1 percent). A small number of respondents indicated that the data are collected but are not used.

The majority of respondents, (73 percent) reported that policies regarding collection of data on race and ethnicity were *not currently undergoing revision*. Six percent said the policies were being revised and 16 percent reported that they did not know. The main revision to policies related to increasing the number of categories patients had to choose from when self-identifying race and ethnicity.

Seventy-two percent of the respondents that collect racial or ethnic data did not see any *drawbacks to collecting such data*. Drawbacks that were reported most often included discomfort on the part of the registrar or admitting clerk asking the patient for the information; problems associated with the accuracy of the data collected; a sense that patients might be insulted or offended, or resist answering questions about their race and ethnicity; patients often don't "fit" the categories that are given; and a fear that data may not be kept confidential. Also mentioned were the possibility that collecting data on race and ethnicity might be used to profile patients and discriminate in the provision of care.

In the case of primary language, we simply asked whether hospitals collected data on a patient's primary language, but did not include detailed follow-up questions in order to keep the length of the survey reasonable.

Forty percent of hospitals collect *data on patients' primary language*, 52 percent do not, 3 percent of respondents reported that they did not know, and 5 percent did not respond to the question.

Hospitals that did *not* collect data on race and ethnicity (*n* = 54) were asked to give reasons why. Sixty-one percent stated that it was *unnecessary* to collect data on patients' race and ethnicity, 18 percent felt there was no reliable system for data collection, 17 percent said there was no good classification system for race or ethnicity, 7 percent said the data were too costly to maintain, 7 percent said the data would be unreliable, 5 percent said it was legally allowed but not authorized by the hospital, and 5 percent stated it was prohibited by law or external regulation.

Only one of the 54 hospitals not currently collecting racial or ethnic data indicated it was engaged in efforts to change its policies related to the collection of these data for the purposes of determining patients' primary language. Respondents from hospitals that were not changing their existing policies stated there was no need to collect data on race and ethnicity because it was unnecessary, it was not a priority or a concern, the patient population was over 95 percent Caucasian, and one stated it was "against the law."

Hospitals that did not collect data on race or ethnicity were also asked whether they saw any *drawbacks* to collecting these data; of these hospitals, 44 percent answered "no" and 56 percent answered "yes." Respondents stated that the time and resources involved in collecting and managing the data would be a barrier to its collection. One thought it was an invasion of privacy. It appears there is concern that providers will use the information to discriminate in the provision of care. Whether this occurs or not, there is concern that patients will *perceive* that care will be different based on their race or ethnicity if the information is provided. One respondent voiced concern that knowledge of a patient's race and ethnicity would lead to "segmenting service delivery, discrimination, and multiple standards of care." Another wrote, "some people feel these questions signify that they will be treated differently from other patients." Others took an almost defensive stance, questioning the need for such data and stating that "all of their patients are treated the same" and asking "does it make a clinical difference?"

It is interesting to note the differences and similarities in the drawbacks identified by hospitals that do and do not collect data on race and ethnicity. Hospitals that did not collect data indicated that the time and resources needed to collect and manage the data were a barrier to collection, whereas hospitals that collect the data did not. Both expressed concern about the possibility that discrimination would result in the provision of care, although hospitals that do *not* collect the data mentioned it considerably more often than hospitals that did. The drawbacks mentioned most frequently by hospitals that *do* collect data were related to staff and patients feeling uncomfortable or offended by questions about race and ethnicity, and concerns about the accuracy of the data, particularly for individuals who do not "fit" the categories provided.

SUMMARY OF SURVEY OF AAHP MEMBER HEALTH PLANS

In February 2003, a survey on the collection of racial and ethnic data was developed collaboratively by the MSU Institute for Health Care Studies and the American Association of Health Plans. In March 2003, the survey was sent to a sample of 158 health plans based on the sample methodology described below. A total of 38 completed surveys were returned.

Sample Methodology

The sample frame was a database from The InterStudy Competitive Edge, Part I: HMO Directory, Version 12.1 (using data as of July 1, 2001), consisting of 500 HMOs. This source was chosen because of its reliability in reporting both enrollment information and types of products offered.

From the HMOs listed in the database, health plans were selected that were AAHP members as of February 1, 2003. From this sublist, AAHP member health plans that had recently been included in a sample for a similar project were removed. The final sample consisted of 170 health plans. In all, 12 plans were deemed ineligible.

The remaining 158 health plans represented 59 million enrollees in HMO, POS, and PPO plans. The final sample of health plans responding to the survey represented approximately 16.3 million enrollees from the various health plan products and provided information about 69 commercial, Medicaid, and Medicare health plan products.

Findings

Of the 38 health plans that responded to the survey, 13, or 34 percent voluntarily collected data on the race and ethnicity of their health plan members and 23, or 61 percent did not at this time. Two other respondents either did not know or did not respond to the question.

Of the 13 health plans that reported voluntarily collecting data, the following reasons were provided for establishing policies or practices to collect such data: six plans reported that they collected data in order to design translation materials or to improve quality, another five plans indicated that they recognized the benefits of data collection or that they use the data for disease management. Other reasons given were that data collection is required by law or regulation (3), the data were used to screen high-risk populations (3), the data were used to help marketing efforts (3), and data collection was permitted by statute (1).

Data collection primarily occurred upon health plan enrollment. Fourteen responses indicated that data on race and ethnicity were voluntarily collected from the state Medicaid or federal Medicare enrollment files (9), directly from the enrollee (4), or by the caretaker or guardian (1).

Among the responses received from health plans that did not voluntarily collect data, eight stated that data collection was unnecessary, seven stated that it was legally allowed but not authorized by the health plan, six stated that either they had concerns about the enrollee's reactions, or there was no good or reliable system for collection, two stated that either it was prohibited by law or external regulation or they did not know, and one response indicated that either the data would be unreliable, too costly to collect and maintain, or there was no good or reliable classification system for race and ethnicity. Fifteen health plans stated that there were drawbacks to collecting these data. Specific issues identified were concerns about enrollees' perceptions, there was no good or reliable system for data collection, the data collection was legally allowed but not authorized by the plan, data would be unreliable, too costly to collect and maintain, or there was no good or reliable system to classify race and ethnicity.

SUMMARY OF SURVEY OF AMGA-MEMBER MEDICAL GROUPS

In early March 2003, an e-mail survey was sent to a designated contact person in each of the 250 member organizations of the American Medical Group Association. The e-mail was sent from the AMGA offices in Alexandria, Virginia, and the cover note described the purpose of the survey as gathering information about current practices and policies with regard to collection of data on race and ethnicity. An e-mail response option was provided; all responses were returned to the AMGA office and automatically entered into a database. A total of 83 valid surveys were returned, and another seven respondents replied by e-mail that they did not collect the information.

Findings

In response to the opening question about whether the group collected data on the race or ethnicity of patients, 21 of the 83 respondents (25 percent) answered "yes." All but one of the others responded "no" and one was "not sure."

The groups that did not collect data gave a variety of reasons for not doing so. The most frequently chosen reason for not collecting data was "concerns about patient reactions" (26/62 or 42 percent of those not collecting data). Other reasons included "unnecessary" (37 percent), "no good/ reliable system for collection" (25 percent), and "data would be unreliable" (18 percent). Only two respondents felt that data collection was prohibited by law or external regulation. (Percentages add to more than 100 percent because respondents could choose more than one option.)

Those groups that did report collecting data on race or ethnicity did so mainly for internal purposes rather than for external reporting. Fifteen of the 21 groups collecting data on race or ethnicity said that they used the data for internal purposes. Ten of the 15 mentioned using the data for quality improvement or disease management purposes. Six of the 21 groups said that they shared data with federal, state, or local government agencies—primarily state or local health departments.

In most instances, the patient was the source of data on race or ethnicity. Seventeen of the 21 medical groups chose "patient self-identifies" in response to a question about who provides data. Caregivers or health care providers were much less frequently mentioned as the source of data. Data were collected most often at the first visit or new patient registration (15 out of 21 groups), but in eight groups the provider's note in the medical record was identified as the point at which data were collected. (Again, groups had the option of choosing more than one answer.)

Summary

Like health plans, medical groups are less likely than hospitals to collect data on patients' race and ethnicity. Those groups that do collect data do so primarily for internal quality improvement or disease management purposes rather than because of an external reporting requirement. Groups that do not collect data feel that it is either unnecessary or potentially troubling to patients, or that it cannot be done reliably.

Appendix G

Racial and Ethnic Data Collection by Health Plans

Carmella Bocchino *

INTRODUCTION

Access to quality health care services is a prominent focus of health care organizations, researchers, and policymakers in America today. For culturally diverse populations, access to quality health care is often hampered by a variety of socioeconomic and cultural factors.

Reports such as *Diverse Communities, Common Concerns: Assessing Health Care Quality for Minority Americans* (Collins et al., 2002), have demonstrated how race, ethnicity, and English proficiency can affect access to quality health care as the United States becomes a more racially and ethnically diverse nation.

Health plans have recognized the importance of responding to patients' varied perspectives, beliefs, and behaviors about health and well-being, as well as the considerable health consequences that will result in a failure to value and manage cultural and communal differences in the populations they serve. Through the emerging field of culturally competent care, health plans are developing strategies to reduce disparities in access to and quality of health care services. The collection of data on race and ethnicity is a first step in designing and advancing such strategies.

Health plans generally are supportive of the collection of racial and ethnic data on their members. They see these data as having great utility in

---

*Carmella Bocchino is vice president for Medical Affairs of America's Health Insurance Plans.

272

a number of areas, which are delineated in this paper. Barriers, however, do exist—collection is not consistent across the industry and often fragmented—which make it difficult to evaluate the quality of such data and subsequently determine solutions to advance culturally competent care. Even so, some strong examples of data collection and related innovative strategies for use are emerging.

To date, several organizations, notably the Agency for Healthcare Research and Quality (AHRQ), the Commonwealth Fund, and the National Quality Forum, have identified many uses for the collection of data on race and ethnicity. They include:

- Understanding the scope of health disparities affecting health plan members and stimulating action.
- Identifying and tracking similarities and differences in performance and quality of care in various geographic, cultural, and ethnic communities.
- Revealing socioeconomic and other demographic characteristics that contribute to differing proportions of disparities.
- Creating and using reports that focus on quality of care issues for minority group patients.
- Understanding etiologic processes and identification of points of intervention.
- Designing targeted quality improvement activities.
- Facilitating the provision of culturally and linguistically appropriate health care.

Health plans view the collection of data on the race and ethnicity of their members as having great utility in a number of areas, such as evaluating the differences in care being received by plan members; designing culturally appropriate educational and other member communications; and implementing clinical and service quality improvement activities.

What follows is a description of interviews conducted across a sample of AAHP's[1] member health plans, assessing their efforts to collect racial and ethnic data; offering examples of some current and potential methods of collection; identifying real and perceived barriers to data collection; and detailing the usefulness of such data for health plan programs.

REPORT ON INTERVIEWS OF AAHP MEMBER HEALTH PLANS

The American Association of Health Plans (AAHP), now known as AHIP, was asked by the National Research Council of the National Acad-

---

[1]AAHP merged with the Health Insurance Association of America in November 2003; the new organization is called the America's Health Insurance Plans (AHIP).

emy of Sciences to provide information to the panel on DHHS Collection of Race and Ethnicity Data. The panel was convened to examine the adequacy of data on race and ethnicity collected or used by DHHS programs, and will issue guidance to the DHHS regarding the following:

- What data on race and ethnicity are private organizations and providers collecting?
- What is the availability and quality of the data collected?
- How can private organizations benefit from the collection of these data?
- How are the data used?
- What are the barriers to collecting these data?

AAHP was asked to evaluate whether and how health plans collect racial and ethnic data. The panel believed there would be value in using an interview method that provided for and encouraged interactive questioning. AAHP conducted telephone interviews with a sample of member health plans across the country to identify and highlight the issues surrounding such data collection and summarize these efforts.

Methodology

A sample of 30 health plans was selected for interviews using a two-step methodology.

Step 1. Sixteen AAHP member health plans were chosen, because these plans were known to have initiated activities related to this project.
Step 2. Fourteen health plans were selected from the pool of health plans responding to the 2002 AAHP Industry Survey (n = 194). A subset was created from these 194 health plans using the following criteria: current AAHP membership and enrollment of at least 100,000 members. A random selection of 14 plans was chosen from the subset.[2]

---

[2]The list of responders to the 2002 AAHP Annual Industry Survey was used as the sampling frame for this project because it contained health plans verified as eligible to participate and had current contact information. The 2002 industry survey sampling methodology used the following selection criteria: five large national plans were sampled with certainty, because to exclude them from the sample would distort the national data, as they represent a large population of health plan members. A subsequent sample of additional plans was selected in a randomly stratified manner as to enrollment: very large (≥6 million members), large (>370,000 but <6 million members), medium (130,000 to <370,000 members), and small (<130,000 members).

RESULTS

Demographics and Composition of Health Plans Interviewed

Of the 30 plans in the sample, 24 (80 percent) completed the interviews. Collectively, the 24 health plans served approximately 49 million people and provided services in all 50 states, Washington, DC, the Commonwealth of Puerto Rico, and the Territory of Guam. The health plans represented approximately 30 percent of the total health plan enrollment in the United States. In terms of enrollment, nearly 63 percent of the health plans interviewed were defined as very large (≥6 million members) or large (>370,000 but <6 million members). An additional 38 percent of the health plans were defined as medium-sized (130,000 to <370,000 members) or small (<130,000 members).

Fifty percent of health plans interviewed offered two lines of business (commercial and Medicare); 42 percent offered three lines of business (commercial, Medicaid, and Medicare); one health plan offered only Medicaid and Medicare; and one plan offered only a commercial product.

The majority of the respondents interviewed for this study were responsible for either quality management/accreditation (50 percent) or market research/sales (21 percent) activities within the plans. Staff responsible for customer service, human resources/strategic planning, data and statistics, and health/disease/case management also participated in these interviews. The majority of respondents interviewed have been with their current health plan for 4 or more years, while less than a quarter of respondents were with health plans fewer than 4 years.

Summary of Findings and Trends

The health plans surveyed uniformly agree that the collection of racial and ethnic data is a part of good business practice. Such data allow for the identification of populations that may benefit from a customized approach to working with providers who can deliver culturally competent care. Health plans also identify information on sex, age, education, and geographic location as additional elements that contribute to the ability to meet perceived and real health care needs. Minority populations identified and served by the health plans interviewed were African American (the largest minority population identified), followed by Hispanic/Latino, and Chinese, Native American, Hmong, Korean, Russian, and Japanese.

While health plans use a number of sources for the collection of racial and ethnic data, including both directly from members and indirectly from other sources, the most commonly used are census data (88 percent). Health plans use these data to match the Zip Codes of their members to Zip Codes within census blocks and then make a determination, based on the demo-graphics of the census block, as to member race or ethnicity. This method is followed in frequency by individual self-identification on member satisfac-tion surveys—e.g., the Consumer Assessment of Health Plans Survey (CAHPS) or other plan-initiated surveys—(79 percent) and Medicare/Med-icaid enrollment files (29 percent). Although 25 percent of health plans ask for race and ethnicity on their enrollment forms, the questions are volun-tary and frequently left unanswered. It is for this reason that most health plans use "indirect methods" for data collection, such as census tracking or questions on primary language spoken as a proxy for racial and ethnic data. Health plans perceive that members may be reluctant to share their race and ethnicity, but that members tend to view information about language as a benign request. Language preferences, at a minimum, indicate how best to communicate to the member about access, services, and health education materials. Additionally, it often provides reasonable insight into ethnicity. Finally, health plans may receive additional information about race and ethnicity from members enrolled in health plan disease management pro-grams. For example, as case managers conduct outreach to individual mem-bers enrolled in disease management programs, the need for special ser-vices, such as interpretation or translation, may be identified.

Information about individual providers (e.g., language capabilities) is frequently used to assist members in provider selection or to "match" members with specific health professionals who can best meet their linguis-tic and cultural needs and thus enhance the quality of care and improve patient outcomes.

At present, health plans are using racial and ethnic data in two major areas: to address preventive care issues within specific populations, and to identify populations at higher risk for certain chronic conditions. Other uses identified by health plans include the ability to apply a "loose evalua-tive" process to determine the consistency with which care is being deliv-ered across different racial and ethnic groups—especially with regard to specific conditions; to assess the "representativeness" of the data collected in member surveys (i.e., how accurately the data reflect the populations served); to design culturally appropriate educational and other member communications; to implement clinical and service quality improvement programs that address the unique needs of racial and ethnic subpopula-tions; and ultimately to serve populations better by identifying and re-sponding to their unique needs.

All of the health plans interviewed indicated that they generally target the most prominent segments of diverse populations within their service areas for personalized health care offerings (e.g., translated materials, interpreters). Health plans use the services of interpreters and translators (e.g., AT&T) as requested or as the need is identified to enhance the quality of health care services and outcomes. These services, however, have major constraints: they are expensive and have limitations; the translators are equipped for conversational translation, not trained in medical terminology; and they are available only by telephone. Translation needs often are augmented through the use of health plan or provider office staff with appropriate language skills. Differences in dialect and place of origin, however, can decrease the accuracy of translation. For example, people from Mexico may use different dialects than individuals from Puerto Rico, although both are categorized as being of Hispanic/Latino descent.

At this time there appears to be no consistent approach to collecting racial and ethnic data within and between health plans, or throughout the health care system. Although frequently such data are recorded as a result of customer service logs or membership interactions, they are seldom shared "systemwide" (for example, consumer service log entries are not integrated into the membership database). Racial and ethnic data are, however, routinely targeted for specific program use, e.g., disease management or quality improvement programs or customer service.

In many cases these findings mirror those of a Commonwealth-HRSA pilot study (Nerenz et al., 2002), which concluded that such data can be obtained through one or more "work-around" methods until more direct methods of data collection are implemented.

Additionally, plans participating in the Medicare+Choice (now called Medicare Advantage) program are required to take part in a special national project on culturally and linguistically appropriate services (CLAS) or clinical health care disparities (CHCD).

Initiated by the Centers for Medicare and Medicaid Services (CMS) in 2003, the CLAS project focuses on language access and organizational support in such areas as providing oral language translation services, assessing the diversity of health plan members and the community, assessing the cultural and linguistic competence of health plan, and developing a diverse workforce. The CHCD project requires Medicare Advantage plans to focus on diabetes, pneumonia, congestive heart failure, or mammography for any one or more of the following populations: American Indians/Alaska Natives, Asians, Black/African Americans, Native Americans/Pacific Islanders, and Hispanic/Latinos.

These projects provide a strong incentive to better understand how to collect such data and, hopefully, will produce some templates for data collection that can be applied throughout the health plan community and the health care industry.

Availability and Quality of Data Collected

Health plans may have more racial and ethnic data available—at least of a general nature—than first recognized. Multiple sources for collection do exist within health plans and in many cases are being used, such as Medicaid/Medicare enrollment files, medical records data, self-reported items in surveys (CAHPS or plan-initiated), and customer service records. Publicly available data and software, such as census data and surname-recognition software, have been purchased and used to enhance existing data sources. Leveraging the race, ethnicity, socioeconomic status (SES), and other public health data in state databases through file linkages with health plan data is an infrequently used strategy due to confidentiality and funding constraints, but is worthy of further consideration. It is generally acknowledged that the difficulty of integrating the information across various databases and systems, as well as the time and human resources required for initiative development, presents major and often insurmountable challenges, especially for smaller health plans.

Health plans also cite their lack of confidence in the quality of self-reported racial and ethnic data. As a result, health plans have found that multiple approaches to the collection of such data may be necessary. Health plans have collected primary language data on their members for years; however, it is an optional field on enrollment forms and the vast majority of members leave it blank. One approach that several health plans have used and many others are considering is the application of geocoding to members' residential address information in health plan enrollment files. Geocoding permits health plans to use census data to create proxy variables for a member's race and ethnicity based on the prevailing characteristics of the census block in which a member resides. Geocoding also can provide information on other important socioeconomic variables that can affect health risks and medical care delivery, such as education and income. While geocoding is not 100 percent accurate, it is reasonably reliable and has been found useful for identifying high-risk members and for identifying potential disparities in care.

Consumer Benefits Derived from the Use of this Information

The health plans interviewed identified specific benefits and applications for the collection of racial and ethnic data. As might be expected, given their emphasis on preventive care, health plans most frequently responded that they use these data for addressing preventive care issues within specific populations and identifying populations at higher risk for chronic conditions. The interviewees also realized that these data would be extremely helpful for evaluating the consistency and "patient centeredness" with which care is delivered and for designing clinical and service quality improvement programs that address the unique needs of racial and ethnic subpopulations. In addition, designing culturally appropriate educational and other member communications (written, voice, and electronic) and developing other effective initiatives were cited. Health plans stated that valid and reliable racial and ethnic data also could be used for assessing the "representativeness" of the data collected in member surveys. Health plans could use these data to determine the racial and ethnic make-up of respondents—as well as nonrespondents—to member surveys, to assess how close their response rates (or nonresponse rates) were compared with the overall demographic of the health plan. Finally, health plans recognized that, ultimately, these data can assist them in better serving their populations by identifying their unique cultural needs.

Identified Barriers to Data Collection

Legal statutes and regulatory mandates, at any level of government, can serve as barriers to the collection of racial and ethic data. For those health plans that operate in several states, multiple considerations increase the level of complexity. Although perceived otherwise by the public and across the health care industry, legal barriers to data collection on the race and ethnicity of their members are generally absent except in four states that have laws or regulations restricting health plans from collecting such data (see the review of both federal and state policies and regulations concerning the collection of racial and ethnic data at the end of this appendix).

Almost 63 percent of the respondents cited legal concerns (perceived or real) as the most frequent barrier to data collection. Negative member reaction/response to such data collection—the perception of potential discrimination, distrust, lack of understanding of the purpose(s) (often exacerbated by language barriers)—and/or simple noncompliance also were stated as barriers to data collection. Seventy-five percent of interviewees indicated that negative responses could be reduced if health plan members better understood the positive aspects of such data collection. Incomplete information from members or lack of confidence in the accuracy of self-reported racial and ethnic data was an additional concern voiced by health plans (Arday et al., 2000).

If these perceived barriers are removed, through educational and informational efforts, the result would, most likely, be helpful in advancing both the collection and use of the data to identify populations and design programs to reduce health care disparities. To have any lasting influence, these efforts must be directed at both the member population and the health care industry in general.

Additional findings suggest that while health plans acknowledge the increased diversity and varying health needs of the populations they serve, "going it alone" in terms of data collection presents some barriers that may be difficult for a single health plan to overcome. Limited resources (financial as well as human) for data collection and for the design and implementation of programs for specific populations were identified by the majority of health plans as a major obstacle. Also, system changes required to capture and share the data require tailoring of any "off-the-shelf" technology—an added expense.

Health plans also cited specific examples of potential implementation barriers to data collection. Plans are discouraged by the "hassle factor" of policies and procedures required by governing agencies (e.g., the state-level Department of Insurance) to revise enrollment forms to collect racial and ethnic data. The Health Insurance Portability and Accountability Act (HIPAA) enrollment transaction form (834) and HIPAA claims/encounter form (837) also were identified as problematic.

Health plans, health care providers, and employers use a variety of methods to complete claims/encounter and enrollment transactions. For electronic claims/encounter transactions, health care providers use the HIPAA 837 form to submit requests for payment and transmit information about health care being provided when there are no direct claims under a provider's contract. For claims submitted on paper, health care providers submit either the Health Care Financing Administration (HCFA [now CMS]) 1500 form to bill for professional services or the HCFA 1450 form to bill for institutional services. For electronic enrollment transactions, employers may use the HIPAA 834 form to transmit enrollment and disenrollment information to a health plan in order to establish or terminate an individual's health insurance coverage. The HIPAA standard enrollment transaction form (834) designates racial and ethnic data as a "situational field," but restricts collection to that done by an employer.

Most claims/encounters are submitted electronically and the HIPAA 837 claims/encounter transaction does not include fields for race and ethnicity. For claims submitted on paper, neither the institutional claim form (HCFA 1450) nor the professional services claim form (HCFA 1500) contains a field for racial and ethnic data. Similarly, health plans are increasingly relying on electronic submission of enrollment data, yet HIPAA requires the use of the standard enrollment transaction (834) only by employers who self-insure under an Employee Retirement Income Security Act (commonly known as ERISA) plan; for others it is voluntary. Many of these employers may continue to submit enrollment data in proprietary electronic formats or on their own unique paper forms, which may or may not include data on race and ethnicity or may not capture and code the data in a consistent manner. For those self-insured employers covered under HIPAA, the Implementation Guide for the 834 enrollment transaction designates race and ethnicity as a "situational" data element, for which collection is dependent both on mutually agreed upon contractual reporting requirements between an employer and health plan *and* on the collection of these data not being prohibited by federal or state regulations. Furthermore, racial and ethnicity data are required only if the enrollee is the subscriber unless the contract between the plan sponsor and the payer requires such reporting for dependents. Until a strong business case for reporting data on race and ethnicity can be made for employers, securing such contractual agreements may prove difficult. Health plans may be reluctant to push the issue with employers if the additional "hassle factor" might put them at a competitive disadvantage relative to a plan that doesn't request such reporting.

Even if the HIPAA standard enrollment transaction were modified to make race and ethnicity required data elements and employers were to voluntarily adopt the modified standard, an enrollment transaction is usually generated only for new members in a health plan option and for those who change or terminate their coverage options. Therefore, collection of racial and ethnic data in the enrollment transaction alone will not provide information on the vast majority of health plan members who have not recently joined the plan or changed their coverage during a given year. Since the large majority of these enrollees are likely to use at least one covered service in a given year, collection of race and ethnicity on the 837 claims encounter transaction would be a viable way to consistently capture race and ethnicity for these members.

As part of any future modifications to the HIPAA standard administrative transactions, some health plans advocate making race and ethnicity a required field on HIPAA standard enrollment transactions, as well as for claims/encounter transactions. Their objective is to achieve uniformity across data transactions, not only in how information is captured but also in how data are represented.

Using Racial and Ethnic Data

· Although there are good data demonstrating disparities in care, there are limited strategies on how to reduce the existing gaps in health care outcomes for diverse populations. As health plans work on strategies to collect racial and ethnic data, they also are assessing how to best utilize the data to improve health outcomes. While we know there are some targeted interventions to improve access through outreach and understanding, unlike disease-specific interventions, effective population-specific interventions have not been identified. Many health plans—while acknowledging the potential value of racial and ethnic data—do not have a specifically defined program in mind for the use of such data.

Our interviews demonstrated that there are many health plans incorporating—some in a more structured form than others—the conclusions of a report entitled *Cultural Competence in Health Care: Emerging Frameworks and Practical Approaches* (Betancourt, Green, and Carillo, 2002). The report outlined three levels of promising practices—organizational, systemic, and clinical—used by health care organizations to increase cultural competence. Practices such as providing on-site interpreters and involving community (member) representatives in quality improvement efforts, as well as integrating cultural competence into training for health care providers, are being discussed and incorporated in increasingly larger numbers into the health plan structure.

Although health plans have implemented many changes in data collection efforts, there is no universal approach to the initiation of racial and ethnic data collection and cultural competence activities. Forty-two percent of the health plans interviewed indicated they had a CEO-level task force/directive; 33 percent stated that racial and ethnic data collection and cultural competence efforts were simply integrated into the "daily activities" of the health plan; 13 percent replied that they used both processes; and 13 percent indicated that they had no "formal" program in place.

As previously stated, some innovative health plan models are emerging that demonstrate the industry's efforts to better understand the unique needs of the members and communities they serve, as well as the sociocultural influences on individual health beliefs and behaviors. The following section offers a few examples.

Generic Examples of Health Plan Efforts to Collect and Use Racial and Ethnic Data

• One health plan is beginning to collect racial and ethnic data from its members via the enrollment form and in its disease management programs. This information is used for population-specific health improvement efforts, such as increasing Pap smear rates and the development of a maternity management program to decrease premature deliveries, as well as specific disease management programs.

• Through racial and ethnic data provided on membership satisfaction surveys, one health plan determined that members of Hmong descent showed lower than average levels of satisfaction. Subsequent interviews with the Hmong members revealed gaps in understanding of the workings/process of the health system and cultural expectations of cures that resulted in frequent switching of providers. Peer education was initiated to communicate Hmong cultural expectations to providers and explanations of the health care system to Hmong members.

• A nationwide health plan is collecting information from indirect sources—such as requests from members for health plan materials in languages other than English, for providers with specific language capabilities, and for physician notes and medical records—to identify and recruit members for focus groups among specific diverse populations that will add to the body of knowledge in the literature on cultural approaches to health care promotion and chronic condition management.

• Another health plan uses the "primary language preference" information obtained from member enrollment forms, combined with efforts to identify members with more common Chinese names, to send Chinese language materials and to inform these members about services available through their plan's Asian Initiative.

• Four large health plans in one market have collaborated with the state department of public health both to link member files with state public health data files on a project-specific basis (e.g., prenatal care and birth outcomes, stage of cancer at diagnosis) and to collaborate on state-funded surveys (e.g., Behavioral Risk Factor Surveillance System [BRFSS], Diabetes Control Program survey). These plans then have been able to assess access to and utilization of services as well as clinical quality measures for member subpopulations using the racial, ethnic, and SES data collected in the state databases and surveys.

CONCLUSIONS

Health plans recognize that cultural diversity and beliefs offer numerous challenges for the health care industry, as well as opportunities. Plans are interested in participating in a coordinated, directed effort to accelerate the collection of accurate racial and ethnic data. But obstacles to achieving this goal do exist. Currently, health plans have collected some racial and ethnic data, mainly through a number of indirect methods. Although these sources frequently are not sufficiently recognized or consistently utilized by health plans, health plans are working to improve access to appropriate, culturally sensitive health care and to decrease health disparities.

Many of the barriers discussed in this paper could be removed through collaborative efforts to simplify, streamline, and coordinate methods of data collection as well as through agreement on racial and ethnic categories. Such efforts must then be coupled with a public education initiative, focused on the great utility and positive health effects that racial and ethnic data collection will provide.

At the core of this effort is instilling (perhaps renewing) a sense of public trust that this information will be used with respect and only for the benefit of health plan members, that is will not place them in any jeopardy. Including champions from designated populations at all stages of racial and ethnic data collection efforts—from inception, to design, to implementation, application, and evaluation—has helped to achieve this goal in several communities. The use of racial and ethnic data in health plans' disease management programs also has demonstrated the benefits of this information—improving the prevention or control of disease through targeted outreach, education, and follow-up.

Regarding specific implementation efforts, while larger organizations have more flexibility in control of the human resources needed for change and to assemble task forces for this purpose, "going it alone" is not a viable option for the majority of health plans. For this reason, our respondents stated that the efforts they believed would be well received and prove most effective would include: regional collaborations; "mentorship" with smaller plans; coordinated public/private efforts toward the education of health plans and the health care industry regarding legal issues; information sharing and confidence building/assurances to the public about the intended and specific use of racial and ethnic data; and standardization of data collection methods.

The collection of racial and ethnic data on the populations served by health plans is an initial step toward the development of a truly culturally competent health care system. Understanding the factors that prevent minorities from obtaining quality health care and how those factors interact with the health care system is key to closing the gap in health care access and outcomes between majority and minority populations. Race and ethnicity data collection offers opportunities for health plans and providers to focus on their members' diverse values, beliefs, and behaviors, and to tailor the structuring of their health care services to meet each patient's social, cultural, and linguistic needs. Through these efforts, reaching the goal of quality health care that is effective, safe, patient-centered, and equitable for all Americans can be achieved.

Recommendations

Based on our findings, we make the following recommendations:

• Development of a coordinated, uniform approach across the health care industry to accelerate the collection of accurate racial and ethnic data, which would include input and active participation from health plans, employers, and federal and state governments. Efforts that would be most effective and well received may include:

– Community-based collaborations that simplify, streamline, and coordinate methods of data collection.

– Mentorship initiatives with "smaller" health plans.

– Development of a strong business case for reporting by employers.

– Public/private efforts to educate the health plans and industry about legal issues.

– Information sharing and confidence building to instill public understanding and trust in proper use of collected data.

– Standardizing data collection methods developed with leadership from the federal government (HIPAA—or similar national approach—would move this effort beyond the type of plan within which an individual may be enrolled).

• Recognition, designation, and support of "champions" from designated populations to lead and guide the collection of racial and ethnic data at all stages.

• Identification of models that work to balance the extensive research concentrating on gaps in health care quality linked to race and ethnicity.

• Funding of new research directed at specific methods of how to reduce or eliminate "gaps" in medical care experienced by some racial and ethnic minorities. This would include identification of specific factors that prevent culturally diverse populations from obtaining quality health care and how these factors interact with the health care system.

Review of Federal Policies and Practices and State Laws Regarding Racial and Ethnic Data Collection

Federal Level

A study by The Commonwealth Fund (Perot and Youdelman, 2001) delineated the context in which health-related data collection and reporting by race, ethnicity, and primary language takes place at the federal level, particularly within the U.S. DHHS. The authors conducted a survey of the statutes, regulations, policies, and procedures of federal agencies to identify when the collection and reporting of such data are required and assessed the interpretation and implementation of existing laws and regulations, as expressed by 60 respondents associated with the administration of health care services.

Four major findings emerged from the investigation:

• Collection and reporting of data on race, ethnicity, and primary language are legal and authorized under Title VI of the Civil Rights Act of 1964. No federal statutes prohibit this collection, although very few require it.

• An increasing number of federal policies emphasize the need for obtaining racial and ethnic data. There is high-level agreement that primary language data should be collected as well.

• General agreement prevails that racial, ethnic, and primary language data are critical to promote health and quality health care for all Americans.

• Despite these findings, federal data collection is not uniform. Data requirements and methods for collection and reporting vary across federal agencies and do not fully reflect consensus on the value of gathering this information (Perot and Youdelman, 2001).

State Level

Four states—(California, Maryland, New Hampshire, and New Jersey)—have laws or regulations barring health plans from collecting data on race and ethnicity. This prohibition has a significant impact on racial and ethnic data collection, as these states have high HMO penetration rates: California, 48.5 percent; Maryland, 27.8 percent; New Hampshire, 25.9 percent; and New Jersey, 27 percent. The regulations are summarized in the following sections:

California[3]

California has a provision in its insurance code that prohibits health insurers from identifying or requesting an applicant's race, color, religion, ancestry, or national origin on an insurance application. *Since managed care organizations (other than Preferred Provider Organizations [PPO]) are not subject to the insurance code, they are not bound by this provision.*

Note: The DMHC has collected information from MCOs to determine how they address any cultural and linguistic barriers faced by their members. The Office of the Patient Advocate surveyed all the MCO chief executive officers in the state concerning their cultural and linguistic access policies. The responses were voluntary and the results were incorporated into a recently published consumer report card. A table showing some of the services HMOs provide for their members in other languages or in American Sign Language, such as interpreter services and written materials, is available at the California Office of the Patient Advocate's Web site at http://www.opa.ca.gov/report_card/.

Maryland[4]

Maryland has a statute that prohibits the collection of certain racial or ethnic data. The statute states that "an insurer . . . may not make an in-

---

[3]Cal. Ins. Code §§ 20, 688.5, 700, 740, 742, 10141, and 12921; Cal. Health and Safety Code § 1341.

[4]Md. Code Ann. [Insurance] §27-501 (c).

quiry about race, color, or national origin in an insurance form, questionnaire, or other manner of requesting general information that relates to an application for insurance."

New Hampshire[5]

The New Hampshire Insurance Department (NHID) regulation provides that "questions of race or color are prohibited" with regard to all "application forms used in connection with an insurance contract, whether or not attached to that contract."

New Jersey[6]

The New Jersey Department of Banking and Insurance (DBI) has certain regulations that prohibit the collection of racial and ethnic data under certain circumstances. Per the regulation, application forms for individual health insurance "shall not include questions that pertain to race, creed, color, national origin or ancestry of the proposed insured." The DBI prohibition is applied only in the narrow realm of insurance application forms and not at any other point of the process of providing coverage.

REFERENCES

Arday, S.L., D.R. Arday, S. Monroe, and J. Zhang
   2000   HCFA's racial and ethnic data: Current accuracy and recent improvements. *Health Care Finance Review* 21(4):107-116.
Betancourt, J.R., A.R. Green, and J.E. Carrillo
   2002   *Cultural Competence in Health Care: Emerging Frameworks and Practical Approaches.* New York: The Commonwealth Fund.
Collins, K.S., D.L. Hughes, M.M. Doty, B.L. Ives, J.N. Edwards, and K. Tenney
   2002   *Diverse Communities, Common Concerns: Assessing Health Care Quality for Minority Americans.* (Findings from the Commonwealth Fund 2001 Health Care Quality Survey.) New York: The Commonwealth Fund.
Nerenz, D.R., V.L. Bonham, R. Green-Weir, C. Joseph, and M. Gunter
   2002   Eliminating racial/ethnic disparities in health care: Can health plans generate reports? *Health Affairs* 21(3):259-263.
Perot, R.T., and M. Youdelman
   2001   *Racial, Ethnic, and Primary Language Data Collection in the Health Care System: An Assessment of Federal Policies and Practices.* New York: The Commonwealth Fund.

---

[5]N.H. Admin. Rules, Ins. 401.01(i) (5).
[6]N.J.A.C. § 11:4-16.7(a)(1); N.J. Stat. §17B:27A-2.

Appendix H

Biographical Sketches

EDWARD B. PERRIN (*Chair*) is a biostatistician and health services researcher, a professor emeritus and former chair of the Department of Health Services and of the Department of Biostatistics at the University of Washington. He is a member of the Institute of Medicine (IOM) and a former member of the Committee on National Statistics of the National Research Council (NRC). He is a former director of the National Center for Health Statistics, DHHS, and past president, member of the Governing Board, and distinguished fellow of the Academy for Health Services Research and Health Policy and fellow of the American Statistical Association. Dr. Perrin has served as chair of the Health Services Research Study Section and as a member of the National Advisory Council for the Agency for Healthcare Research and Quality, DHHS, and as chair of the Scientific Advisory Committee of the Medical Outcomes Trust. He has served on a number of IOM and NRC study panels, including the Committee to Design a Strategy for Quality Review and Assurance in Medicare, the Committee on Health Care Reform, the Committee on U.S. Physician Supply, the Panel on the National Health Care Survey and the Panel on Performance Measures and Data for Public Health Performance Partnership Grants, serving as chair of the latter two. Dr Perrin's recent research and teaching interests and scientific publications have focused on the development of new methodologies for the measurement of health outcomes and the structure and use of large health data systems in decision making and policy development. He received a B.A. in mathematics from Middlebury College, an M.A. in mathematical statistics from Columbia University, and a Ph.D. in statistics from Stanford University.

HECTOR BALCAZAR is professor of Latino Public Health and chair of the Department of Social and Behavioral Sciences, School of Public Health, Health Sciences Center, University of North Texas. He is also the director of the Center for Cross-Cultural and Community Health Research of the Institute for Public Health Research, School of Public Health. Dr. Balcazar specializes in the study of public health problems of Latinos/Mexican Americans. Dr. Balcazar is a bilingual, bicultural family and public health scientist who has conducted numerous studies of Latino birth outcomes, acculturation and health-related behaviors, cardiovascular disease prevention programs in Latinos, and border health issues. His most recent funded projects include: The North Texas Salud Para Corazon Outreach Initiative; a study on the use of perinatal, infant, and childhood health services among high-risk Mexican American subgroups; the development of a strategic plan for the Latino National Health Collaborative; and a clinical study for Hispanic diabetic patients. As a Latino health specialist Dr. Balcazar provides consultation and leadership to local and national health organizations.

ANTHONY D'ANGELO worked as a statistician, operations research analyst, and mathematician for federal government agencies for 33 years. He developed information systems and analytical models for product assurance, communication systems, and health services. He worked for the Indian Health Service the last 22 years of his career. Mr. D'Angelo served as the principal statistician of the Indian Health Service and manager of the program statistics team. He developed and managed statistical information systems in order to provide data for measuring health status and appraising program activities. He published morbidity and demographic data concerning the American Indian and Alaska Native population. He retired from the government in December 1999 and moved from the Washington, D.C., area to Phoenix, Arizona. He now provides statistical consulting services, specializing in American Indian and Alaska Native data, and volunteers at the World Affairs Council and the Heard Indian Museum.

JOSÉ J. ESCARCE is a senior natural scientist at RAND. Dr. Escarce graduated from Princeton University, earned a master's degree in physics from Harvard University, obtained his medical degree and doctorate in health economics from the University of Pennsylvania, and completed his residency at Stanford University. Dr. Escarce has served on the Health Services Research Study Section at the Agency for Health Care Policy and Research and on the National Advisory Council for Health Care Policy,

Research, and Evaluation of the Department of Health and Human Services. He also serves on the National Advisory Committee of the Robert Wood Johnson Foundation Minority Medical Faculty Development Program and is past chair of the Health Economics Committee of the American Public Health Association. He was member of the IOM Committee on Understanding and Eliminating Racial and Ethnic Disparities in Health Care. Dr. Escarce's research interests include provider and patient behavior under economic incentives, access to care, racial and ethnic disparities in care, and the impact of managed care on cost and quality.

WILLIAM D. KALSBEEK is professor of biostatistics and director of the Survey Research Unit at the University of North Carolina-Chapel Hill. His prior experience includes statistical research with the Office of Research and Methodology at the National Center for Health Statistics and at the Sampling Research and Design Center at the Research Triangle Institute in North Carolina. He is a fellow of the American Statistical Association and a member of the American Association of Public Opinion Research, and the American Public Health Association. He received his M.P.H. and Ph.D. degrees in biostatistics from the University of Michigan. Dr. Kalsbeek's research interests and areas of expertise are in biostatistics, survey design and research, spinal cord injuries, and assessment; he is well known for his work in survey methods.

GEORGE KAPLAN is professor and chair of the Department of Epidemiology in the School of Public Health, senior research scientist at the Institute for Social Research, and director of the Michigan Initiative on Inequalities in Health. He is also a docent at the University of Kuopio in Finland and an associate in the Population Health Program of the Canadian Institute for Advanced Research. He was formerly chief of the Human Population Laboratory and directed the ongoing Alameda County Study. Upon arriving at the University of Michigan, Professor Kaplan initiated the Michigan Initiative on Inequalities in Health (MIIH). The MIIH is a cross-campus initiative with an expanding agenda that will eventually involve all segments of the university community. The mission of the MIIH is to promote inquiry into the causes and consequences of societal inequalities in health and potential remedies, via an interdisciplinary program of discussion, research, and teaching. The goals are to catalyze intellectual interchange, research, and teaching focused on issues raised by the study of inequalities in health through a program of in-depth discussion, lectures, and symposia; an ongoing Forum on Inequalities in Health; small grants; and student stipends. Dr. Kaplan is also director of the Michigan Interdisciplinary Center on Social Inequalities, Mind, and Body (MICSIMB).

DENISE LOVE is the executive director of the National Association of Health Data Organizations (NAHDO), a nonprofit membership and educational organization dedicated to strengthening the nation's health information system. Ms. Love is actively involved in national standards forums, developing integrated Internet systems, and establishing analytic frameworks for major health data sets. NAHDO works to assure the public availability of valid and useful information for diverse audiences. To that end, NAHDO provides technical assistance to states and develops national forums for state, federal, and private-sector leaders. Formerly the director of the Office of Health Data Analysis, Utah Department of Health, Ms. Love's office established statewide reporting systems for inpatient, ambulatory surgery, and emergency department data and in 1996 established Utah's HEDIS reporting for commercial and Medicaid managed care plans. In 1993, her office created the Utah Internet Query System—a Web-based interactive data dissemination tool. An adjunct faculty member in the Department of Family and Preventive Medicine, University of Utah School of Medicine, Ms. Love serves on committees and boards including: Utah Health Insight Governing Council and the Steering Committee of the Public Health Data Standards Consortium. She serves as an adviser and speaker on state health data issues and is the coauthor of several journal articles and numerous state publications. Ms. Love has a bachelor's of science in nursing and a master's of business in health care administration.

JOHN LUMPKIN has been Illinois's public health director since 1991 and is the first African American to hold the position. He has served the third longest tenure of any director since the present agency structure was created in 1917. As director of the Illinois Department of Public Health (IDPH), Dr. Lumpkin oversees an agency of 1,300 employees located in Springfield, Chicago, seven regional offices, and three laboratories who share primary responsibility for the quality of life in the state. IDPH leads a public health system that encompasses partners in various fields and locations, including local and federal health departments, nursing homes, hospitals, physicians, paramedics, plumbers, and food service workers to name a few, who work to promote the health and safety of Illinois's citizens and visitors to the state through the prevention and control of diseases and injury. Dr. Lumpkin's career in public health began with his appointment in 1985 as associate director of IDPH's Office of Health Care Regulations, which oversees the licensing, inspection, and certification of health care facilities. In September 1990, he was named the department's acting director by Governor James R. Thompson. Four months later, newly elected Governor Jim Edgar appointed him to the director's job. Dr. Lumpkin was reappointed to the post by Governor George H. Ryan in January 1999.

ALVIN T. ONAKA is chief of the Office of Health Status Monitoring and state registrar of Vital Statistics for the Hawaii State Department of Health. Prior to returning home to Hawaii he worked for the U.S. National Academy of Sciences and was assigned to the Department of Epidemiology and Statistics at the Radiation Effects Research Foundation in Hiroshima, Japan. From 1974 to 1978 he worked for the U.S. Agency for International Development in Washington, D.C., where he managed population and health projects in Latin American, Africa, and Asia. Dr. Onaka received his Ph.D. degree in demography in 1975 from the University of Massachusetts at Amherst where he was a Population Council Fellow in Demography. He has done postgraduate work in epidemiology and has been on the faculty of the Population Studies Program at the University of Hawaii since 1982. He has served as chair of the U.S. Death Certificate Revision Committee and is an adviser to the National Death Index of the National Center for Health Statistics. Currently he is president of the National Association for Public Health Statistics and Information Systems headquartered in Washington, D.C.

NEIL R. POWE is professor of medicine in the Department of Medicine at the Johns Hopkins University School of Medicine and director of the Welch Center for Prevention, Epidemiology and Clinical Research, an interdisciplinary research and training center at the Johns Hopkins Medical Institutions focused on clinical and population-based research. He also is professor of epidemiology and health policy and management at the Johns Hopkins University Bloomberg School of Public Health, where he directs the Clinical Epidemiology Program and the Johns Hopkins Evidence-Based Practice Center. Dr. Powe's research has also involved clinical epidemiology, health services research, and patient outcomes research in renal and cardiovascular disease. His research has used prospective methods of randomized controlled trials and cohort studies, cost-effectiveness analysis, meta-analysis, retrospective analyses of administrative databases, and survey research. Dr. Powe has published more than 160 articles on these topics. Dr. Powe is principal investigator of the CHOICE study, an End-Stage Renal Disease (ESRD) Patient Outcomes Research Team funded by the Agency for Healthcare Research and Quality and the ESRD Quality (EQUAL) Study. He has extensive experience in developing and measuring outcomes in chronic kidney disease patients using data from prospective studies, the United States Renal Data System (USRDS), Medicare records, and patient surveys. Dr. Powe was a member of the Institute of Medicine Committee on Measuring, Managing and Improving Quality of Care in the ESRD Treatment Setting. He received his MD degree from Harvard Medical School, his MPH degree from the Harvard School of Public Health, and his MBA in health care from the Wharton Graduate School of the University of Pennsylvania. He completed his internship and residency in internal medicine at the Hospital of the University of Pennsylvania. He was also a Robert Wood Johnson Clinical Scholar and fellow in the Division of General Internal Medicine at the University of Pennsylvania. Dr. Powe is a member of the American Society of Clinical Investigation and a Fellow of the American College of Physicians. He is the recipient of several national awards including the best article of the year (2000) by the Academy for Health Services Research and Health Policy and one of the leading African-American physicians in the U.S. (2001) by *Black Enterprise* magazine.

JONATHAN SKINNER is the John French Professor of Economics, Dartmouth College, and a professor in the Department of Community and Family Medicine, Dartmouth Medical School. He has been a research associate at the National Bureau of Economic Research since 1989 and is currently the editor of the *Journal of Human Resources*. He was a recipient of the TIAA-CREF Paul A. Samuelson Award of Excellence in 1996. Professor Skinner's research has focused on the economics of saving and consumption, the effectiveness of medical technology, and issues of fairness and equity in health care.

L. CARL VOLPE is vice president for Strategic Health Partnerships in the Healthcare Quality Assurance Division at WellPoint Health Networks Inc., where he is responsible for developing and implementing strategic projects with hospitals physicians, foundations, and academic institutions to improve services and quality of care for WellPoint members. Previously, as vice president for Health Policy and Analysis, he was responsible for strategic policy development—working within the corporation to anticipate and successfully respond to the potential business impact of state and federal health legislative action. Volpe joined WellPoint in June 1996 as director of Health Policy and Analysis. Prior to WellPoint, he served as director of Health Legislation at the National Governors' Association in Washington, D.C. There, he coordinated the development and implementation of Governors' healthcare policies and advocated those policies to congress and the administration. Previously, he held several positions at the Texas Department of Human Services, most recently as director of health policy initiatives in the Medicaid program. Volpe has experience in a variety of health care arenas including managed care, commercial health insurance, health care financing systems, state/federal relationships in public and private health reform, and Medicaid. He received his Ph.D. in experimental psychology from Northeastern University in Boston. He was awarded a postdoctoral fellowship in applied psychology and evaluation at St. Louis University and was a senior research associate at St. Louis University's Center for Urban Programs.

DAVID WILLIAMS is a professor of sociology and senior research scientist in the Institute for Social Research at the University of Michigan. His previous academic appointment was at Yale University. Dr. Williams's research has focused on differences in socioeconomic status in health in general and the health of the African American population in particular. He has served as a consultant to numerous federal health agencies and private organizations. He has also served on the National Committee on Vital and Health Statistics and chaired its Subcommittee on Minority and Other Special Populations. He is a member of the National Science Foundation's Board of Overseers for the General Social Survey. He has an M.P.H. from Loma Linda University, an M.Div. from Andrews University, and a Ph.D. in sociology from the University of Michigan. He has served on the NRC Panel on Needle Exchange and Bleach Distribution Programs and the IOM Committee on Understanding and Eliminating Racial and Ethnic Disparities in Health Care.

ALAN ZASLAVSKY is an associate professor of Statistics in the Department of Health Care Policy at Harvard Medical School. Dr. Zaslavsky's methodological research interests include surveys, census methodology, missing data, categorical data, hierarchical modeling, small-area estimation, and applied Bayesian methodology. Dr. Zaslavsky's health services research focuses primarily on developing methodology for quality measurement of health plans and other units and understanding the implications of these quality measurements. A large part of his work has been related to the development, implementation, and analysis of the Consumer Assessments of Health Plans Study (CAHPS), a comprehensive program for survey measurement of enrollee reports and ratings of their health plans, and particularly the implementation of this survey for the Medicare managed care population. He has also studied disparities in health care, as indicated by clinical quality measures and utilization. Current work in this area includes involvement in the NCI-funded CanCORS project for evaluation of disparities in processes and outcomes of cancer care.